The Shadows of Consumption

The Shadows of Consumption

Consequences for the Global Environment

Peter Dauvergne

The MIT Press
Cambridge, Massachusetts
London, England

MIT Press books may be purchased at special quantity discounts for business or sales promotional use. For information, please email special_sales@mitpress.mit.edu or write to Special Sales Department, The MIT Press, 55 Hayward Street, Cambridge, MA 02142.

This book was set in Sabon by SNP Best-set Typesetter Ltd., Hong Kong. Printed and bound in the United States of America.

Library of Congress Cataloging-in-Publication Data

Dauvergne, Peter.
  The shadows of consumption : consequences for the global environment / Peter Dauvergne.
      p. cm.
  Includes bibliographical references and index.
  ISBN 978-0-262-04246-8 (hbk. : alk. paper) 1. Consumption (Economics)–Environmental aspects. 2. Environmentalism.
I. Title.
  HC79.C6D38   2008
  333.7–dc22
                                                                2008017002

10 9 8 7 6 5 4 3 2 1

For my children

Contents

Abbreviations

AAMA	American Automobile Manufacturers Association
AMA	Automobile Manufacturers Association (U.S.)
BPA	bisphenol A
BSH	Bosch und Siemens Hausgeräte (German-based company)
CFC	chlorofluorocarbon
DOE	Department of Energy (U.S.)
ELV	end-of-life vehicle
EPA	Environmental Protection Agency (U.S.)
FAO	Food and Agriculture Organization (United Nations)
FSC	Forest Stewardship Council
GATT	General Agreement on Tariffs and Trade
HCFC	hydrochlorofluorocarbon
HFC	hydrofluorocarbon
IFAW	International Fund for Animal Welfare
IMF	International Monetary Fund
IPCC	Intergovernmental Panel on Climate Change (United Nations)
ISO	International Organization for Standardization
LRP	lead replacement petrol
MECA	Manufacturers of Emission Controls Association
MSC	Marine Stewardship Council
MVMA	Motor Vehicle Manufacturers Association (U.S.)
NAFTA	North American Free Trade Agreement
NOAA	National Oceanic and Atmospheric Administration (U.S.)
PBB	polybrominated biphenyl
PBDE	polybrominated diphenyl ether
PCB	polychlorinated biphenyl
PFOA	perfluorooctanoic acid
PFOS	perfluorooctanyl sulfonate

SPCA	Society for the Prevention of Cruelty to Animals
TEL	tetraethyl lead
3M	Minnesota Mining and Manufacturing (before 2002)
UNCHS	United Nations Centre for Human Settlements
UNCTAD	United Nations Conference on Trade and Development
UNEP	United Nations Environment Programme
UNFPA	United Nations Population Fund
USDA	United States Department of Agriculture
WEEE	Waste Electrical and Electronic Equipment (EU directive)
WHO	World Health Organization
WWF	World Wildlife Fund / World Wide Fund for Nature

Preface

Individually, our everyday choices might seem to have no consequences at all for the global environment. What, for example, is the impact of a farmer burning an oil lamp on the Pampas in 1814 or a child eating a bowl of rice in Shanghai in 1889? What, really, is the impact of a man driving a Model T Ford over the Golden Gate Bridge in 1937, or a lawyer switching on a bedside lamp to read *The Grapes of Wrath* in London in 1946, or a teenager opening a Westinghouse refrigerator for a glass of milk in Canberra in 2008? Yet, cumulatively, all these individual acts of consumption—like raindrops in a typhoon—must have consequences. So, too, must all the processes that make consumption possible.

What, then, are the environmental consequences of consumption? How do they affect our health and safety? These may seem like obvious questions in a world of rising consumption and escalating strains on so many ecosystems. Yet few books have ever tried to answer them, and most of these have focused on the more immediate impacts of consumption on local ecosystems and lifestyles. Examining these questions through a wider lens and from a different angle, I analyze not only the direct consequences of consuming, but also the environmental spillovers from the corporate, trade, and financing chains that supply and replace consumer goods: what, to capture the full resulting *global* patterns of harm, I call the "ecological shadows of consumption." Taking this approach puts the primary responsibility for global environmental damage squarely on those with power and wealth while still accounting for the micro-responsibilities all of us bear as consumers. By emphasizing how political and economic processes displace the costs of consumer goods onto distant ecosystems, communities, and times, it reveals the far-reaching effects of our personal choices, effects that few of us ever see—or want to see. And by pushing the analysis beyond progress in improving particular products, it uncovers a core cause of the continuing slide into a full-blown global environmental crisis.

Mapping the pathways of cause and effect for every ecological shadow of consumption would require many lifetimes of research. Rather than lightly touch upon an endless number of products, I chart the histories of just five, representing a range of political economies, from high-end manufacturing to low-end hunting. Each history unfolds over three chapters and concludes with a summary of its lessons for understanding how and why ecological shadows form, shift, and fade.

I begin with two of the most widespread—and deadly—manufactured products of the last 100 years: the automobile and leaded gasoline. Extending my analysis into households, I then turn to a product more often praised for enhancing food security than for causing global harm: the refrigerator. I end with two animal products: beef and the harp seal. Although the last one may strike some readers as out of place (because luxury furs involve far fewer global consequences than beef), I chose to include it to probe how consumption at the periphery of the world economy differs from consumption at the core—essential for a comprehensive analysis, given the large number of consumer goods whose political economies more closely resemble those of the harp seal than of the automobile, leaded gasoline, the refrigerator, or beef.

At every turn, I've sought out sources and examples of positive change—safer cars, cleaner air, superior refrigerators, organic beef, rebounding ecosystems—indeed, far more often than some readers might expect in a book titled *The Shadows of Consumption.* I've done so partly to avoid prejudging my conclusions, and partly to find ways to mitigate the shadow effects of consumption.

This approach has uncovered many trends toward a growing environmentalism over the last four decades. Governments around the world are empowering environmental agencies, reforming national policies, and negotiating and strengthening international environmental laws. International institutions and aid donors are supplying developing countries more funds and technical assistance for environmental initiatives. Global activists are running more campaigns to educate consumers and to lobby firms and governments for further environmental reforms. More corporations are producing more goods under policies of corporate social responsibility. And more consumers are buying more "green" products from "green" markets.

In every case, the globalization of environmentalism is making resource use more efficient by reducing the per unit ecological impacts of consumer goods. Yet these same cases, I will argue, also reveal why so many of the current efforts to manage the global environment are failing. Much of this "progress" is incremental and local, doing more to protect fragments of privilege and power than ecosystems or poor people. Meanwhile, some parts of the earth—places like Africa and the Arctic—are having to pay a disproportionate share of the costs of rising consumption as the globalization of corporations, trade, and financing shifts, intensifies, and casts ecological shadows into more remote regions.

On an ever-larger scale, this accelerating process of change is shifting environmental burdens to fragile ecosystems and to poorer people less able to cope with the consequences. It is hiding costs in distant lands and assigning them to future generations, leaving firms, governments, and wealthy consumers unaccountable for much of the global environmental change now happening. It is contributing to wasteful and excessive consumption among the wealthy while sacrificing basic needs like food and shelter within poorer communities. It is deferring costs and exposing all consumers to long-term health and safety risks. And finally, over time, as states work to protect citizens and economies, it is deflecting environmental costs into spaces with less political and economic power—the tropical rainforests, the poorest communities, the weakest states, the open oceans, the atmosphere—tipping these into crisis and toward collapse.

The global picture shows the continuing decline in the integrity and stability of many of the world's environments. Glaciers and old-growth tropical rainforests continue to recede. Deserts and dangerous chemicals continue to spread. Natural resources and freshwater supplies continue to dwindle. Species continue to die out and oceans to empty of life. And, perhaps most alarming of all, the climate is changing as greenhouse gases warm the earth, with much higher seas and fiercer storms now on the horizon of this century. At the same time, billions of people are at greater risk of environmental diseases and accidents as pollution and congestion worsen in cities, industrial farming expands in rural areas, and chemicals from "new and improved" consumer goods leach into aquifers, food chains, and households.

Most in power no longer publicly dispute such trends. Instead, they tend to point to advances under the rubric of "environmental management"—to business partnerships, eco-efficiency, corporate social responsibility, voluntary compliance—measuring change in their small worlds and applauding policies and institutions that displace costs onto people with less capacity to protest. This creates illusions of progress, leaving far too many of us optimistic about the value of incremental solutions, and far too few of us willing to challenge an "unbalanced world," where so many ecosystems and people are "dying of consumption."

The conclusions of this book strike hard at how the policy community is analyzing and handling global environmental change. Its title, *The Shadows of Consumption*, may give some readers pause, but, as an optimist, I hope that its subtitle, *Consequences for the Global Environment*, will invite them and others to learn how and why consumption has unbalanced our global environment, a necessary step for moving toward a more balanced—and thus a more sustainable—global political economy of consumption: the theme of the final chapter.

Acknowledgments

I am grateful to the Canada Research Chair Program and the Social Sciences and Humanities Research Council of Canada for funding; to my graduate student Ashley Hamilton for tirelessly collecting research for this book; to my graduate students Colin Trehearne and Kate Neville for fact-checking and to Sharon Goad and Josh Gordon for research assistance; and to the staff and faculty in the Dean's Office of the Faculty of Arts at the University of British Columbia—especially Della Krueger—for protecting my author self from my associate-dean self.

I benefited as well from hundreds of conversations with students, friends, and colleagues. My students at the University of British Columbia kept me grounded in the real world and showed a remarkable tolerance for my overenthusiastic approach to research. My special thanks to my graduate students Paula Barrios, Shane Barter, Kaija Belfry, François de Soete, Katherine Hall, Tracey Janes, Lindsay Johnson, Meidad Kissinger, Samantha Kohn, Talusier Arbour LaSalle, Jane Lister, David Seekings, Nicolas Sternsdorff, and Hamish van der Ven.

My thanks as well to the discussants and panelists at the International Studies Association conferences in Hawaii in 2005, San Diego in 2006, and Chicago in 2007 and at the Law and Society Conference in Berlin in 2007, where I presented portions of this book, not just for their helpful comments, but also for their encouragement when the project seemed beyond my reach. I am also grateful to Joe Bowersox for inviting me to Willamette University in November 2007 to present my arguments one last time (leaving me eagerly scrawling out "final" changes).

My years as editor of the journal *Global Environmental Politics* (2001–2008), where I had the privilege of working with the talented scholars Jennifer Clapp, Ken Conca, Beth DeSombre, Karen Litfin, Marian Miller, Mat Paterson, and Paul Wapner, kept me on a steep learning curve throughout. Many others influenced my thinking, too, although no one more than my father, John Dauvergne, who after many decades of conversations somehow continues to bring ever more balance to my understanding of global change. I am especially grateful to Clay Morgan at the MIT Press for his unwavering support and helpful advice throughout and to the anonymous reviewers for the press, whose constructive criticisms gave me the confidence to develop a more hard-hitting message.

By far my most profound thanks go to my partner and wife, Catherine, who had to rely on magic to write her own books in the fragments of time left over after supporting my daily wanderings.

Finally, I dedicate this book to my children, Duncan, Nina, and Hugh, who anchor me in joy in an unbalanced world.

Introduction
The Ecological Shadows of Rising Consumption

1

An Unbalanced Global Political Economy

For thousands of years, the Ayles ice shelf sat off the northern coast of Ellesmere Island, a desolate stretch of glaciers and rock 500 miles (800 kilometers) south of the North Pole. Then, on an August afternoon in 2005, a mass of ice the size of 11,000 football fields suddenly broke free. No humans were nearby to bear witness. But it sent tremors flickering across earthquake monitors 150 miles (250 kilometers) away, and satellites recorded the image of the shelf floating out to sea.

Why did this happen? Was the collapse simply a normal process of nature?

Most scientists think not. They believe that climate change was at least partly responsible. There's compelling evidence to support this. The five warmest years on record are all since 1998. The 1990s was the warmest decade and 2005 was the warmest year in over a century. Over the last few years, hundreds of thousands of square miles of Arctic sea ice have melted under these warmer conditions. If current trends continue, by the end of this century, the North Pole's ice cap could virtually disappear during the late summer season.

Warmer days and nights in the Arctic would seem like a reasonable explanation for the collapse of the Ayles ice shelf. But this raises the underlying question: Why are temperatures rising? Because, the usual scientific answer goes, the amounts of greenhouse gases from human sources (especially carbon dioxide, methane, and nitrous oxide) are rising. World output of carbon dioxide from just consuming fossil fuels currently exceeds 27 billion metric tons a year, up from 18 billion in 1980, and now equal to some 4 metric tons—about the weight of two Hummers—for every man, woman, and child on earth.[1]

Again, however, this answer gives rise to a more fundamental question: Why are these amounts rising?

The direct causes of greenhouse gases span thousands of activities. Factories and furnaces contribute. So do automobiles and airplanes, and so does cultivating rice and raising cattle. Other environmental changes, such as deforestation, desertification, and ozone depletion, are adding to greenhouse gases, too. Just about every act of producing and consuming everywhere seems to contribute. What's more, the total number of people and their per capita rates of consumption continue to rise as well, with environmental impacts far beyond "just" climate change.

Rising Consumption

"Ah," children's storyteller Eric Carle playfully writes, "what we could learn—even if just a little—from the gentle sloth who slowly, slowly, slowly crawls along a branch of a tree, eats a little, sleeps a lot, and lives in peace."[2] Few of the 6.7 billion humans now on earth are so tame (or slothful). The second half of the twentieth century saw our global population grow by 3.5 billion people—a rate of increase faster than in all of recorded history. The global economy expanded at an even faster rate, allowing a per capita increase in gross domestic product (GDP) of 20 percent per decade, or about $3,000 overall from 1960 to 2002.[3] Every month, our industrious and prosperous species continues to increase on average by a little over 6 million members—equal to adding a major city or two. By the middle of this century, assuming past trends hold, our population will exceed 9 billion. Ninety-six percent of this growth will occur in developing countries, with about half in just six countries: Bangladesh, China, India, Indonesia, Nigeria, and Pakistan.[4]

A global hurricane of consumption from these rising populations is gathering force as it sweeps through each generation. For more than 50 years now, per capita consumption of natural resources such as wood, fish, and water has been rising much faster than population growth. The rapid growth in consumption over this time is seen in many statistics. For example, private consumption expenditures (the amount households spend on goods and services) increased more than fourfold from 1960 to 2000, even though the global population only doubled during this period. The future will bring even higher per capita rates of consumption as the developing world pursues the lifestyles of North America and Europe. It has much ground to cover: North America and Europe, with less than 12 percent of the world's population, account for over 60 percent of total private consumption expenditures.[5] China is in hot pursuit, however, with consumption rising in just about every sector.

The Political Economy of Consumption

People buy things for many reasons: need, habit, belief, desire, fear. Most wealthy consumers are free to choose from among many products. Even so, the global political economy determines the "options" as well as guides the collective "choices" of consumers. This is not a static structure, but a shifting set of forces arising from the interaction of many factors along a lengthy chain, from extraction to production to retailing to disposal. The globalization of trade, corporations, and financing is at the core of this global political economy. But new technologies, advertising, and culture shape it, while government policies, activist networks, and global institutions guide it.

Mitigating the environmental impact of this global political economy of rising consumption is one of the biggest governance challenges of the twenty-first century, if not the biggest. Doing so will require a far better understanding of how, why, and to what extent consumption contributes to global environmental change. Chapter 1 begins this trek by unpacking some of the consequences of the globalization of corporations, investment, and trade. Economic globalization is producing many benefits, not only for societies, but also for environmental management. Yet this same process is making it easier and easier for powerful states and firms to deflect the costs of producing, using, and replacing consumer goods into distant ecosystems and onto people at the margins of the global economy. The net result is an unbalanced process of change—unequal within societies, uneven across countries, and unsound in terms of what growing economies draw down from nature—one that, as chapter 2 will show, is casting a disproportionate share of the ecological shadows of rising consumption onto the world's most vulnerable ecosystems, poorest people, and future generations.[6]

The Globalization of Ecological Shadows

As the volumes of trade, investment, and financing and the numbers of consumers continue to rise in a globalizing economy, the ecological shadows of consumption crisscross more and more of the planet.[7] These global patterns of harm arise when states and firms pursuing economic growth, profits, financial stability, and local interests displace the environmental costs of producing, transporting, using, and replacing consumer goods.

More specifically, they arise when multinational companies from countries like Japan and the United States import timber or beef from the tropical rainforests of Southeast Asia and South America, when wealthy consumers in Europe and North America ship used computers to China for recycling, and when countries like China and India spur the growth of their economies without accounting for the costs to the atmosphere or open oceans. They arise as well when companies introduce products without concern for the long-term effects on the health of people or the stability of environments and when states allow products banned as unsafe at home to be exported abroad.

Globalization is accelerating many of the processes casting ecological shadows by integrating—as well as restructuring—economies, institutions, and societies. Many forces are driving it. The continuing spread of capitalism and Western values, which began long ago under modernization and colonization, plays a role. So do faster technologies, such as airplanes, TVs, and computers, by providing efficient and inexpensive transmission belts for people, resources, money, and knowledge. All of which leads more and more to the world becoming "a single place," where changes in faraway lands affect people everywhere with greater speed, force, and frequency, and where borders are ever easier to cross for money, technologies, ideas, and tourists, although, tellingly, not for many of the poor, despite international agreements like the Convention Relating to the Status of Refugees.[8]

Globalization carries with it underlying values and assumptions about how best to organize the world order, which explains why some see it more as an ideology than a set of processes. One core assumption is that indefinite economic growth is possible *and* necessary—and, moreover, that "emerging" economies should follow the path of industrial development and intensive agriculture to ensure ever more consumption, and thus prosperity and stability. Consuming more per capita is a sign that all is well, even when the distribution of its benefits is grossly uneven. The institutional "solution" to such inequalities, which some call "pockets of poverty," boils down to a rather simple formula: rely on the globalization of investment, trade, technologies, and (when necessary) regulations to produce even more goods and services more efficiently—that is, with less labor, resources, time, waste, and environmental impacts. This formula can produce many economic benefits. It can create jobs, it can increase incomes, and it can churn out an abundance of consumer goods and services. It can also enhance environmental management.

Globalization of Environmental Management

Rising per capita incomes over the last half century have been crucial for the emergence of the concept of sustainable development, most commonly defined as "development that meets the needs of the present without compromising the ability of future generations to meet their own needs."[9] Citizens in wealthier places began to demand cleaner and safer environments; activists began to call for action; and, after first resisting, governments began to respond by starting to "manage" the environment, drawing on higher tax revenues to enhance their capacity to regulate. Firms then responded to these consumer and political pressures by developing codes of conduct, expanding environmental markets, and, most significantly, making products more efficiently, with less damage over a life cycle. At the same time, states were able to negotiate hundreds of international environmental agreements, including a few to finance initiatives in developing countries to protect the global environment. Together, such efforts, for example, have helped protect biodiversity in regions having little political or economic power: lichens in Antarctica and elephants in Africa, to list just two.

Today, just about every state is managing the global environment in ways that do not impede economic growth or deter multinational investors, trade, and financing. Most see this as necessary to maintain political and social stability as well as implement environmental regulations (by, for example, hiring staff or buying equipment). This partly explains why international organizations like the World Bank and the International Monetary Fund (IMF) continue to push so many developing states to liberalize investment and trade rules. Given the current world economy, most states and international organizations work hard to avoid currency crashes and capital flight, which can cause not only social but also environmental havoc, as was the case in Indonesia after the Asian financial crisis of 1997–99.

Economic globalization can advance global environmental management in other ways, too. Some multinational corporations raise standards in developing countries by following codes of conduct that are stricter than local laws require—what political scientist Ronie Garcia-Johnson has termed "exporting environmentalism."[10] This can happen because multinationals rely on more sophisticated technologies and management techniques, because social and market forces are pressuring them to go beyond compliance, or because they wish to avoid lawsuits and consumer backlashes. And it can happen because firms are competing for trade or market advantages, or because they have adopted an industry code (such as the chemical industry's Responsible Care), international standards (such as those of the International Organization for Standardization), or internal policies of social responsibility.

At the same time, trade can stimulate more efficient production and create incentives to transfer environmental technologies. It can also encourage producers with comparatively low standards to raise these to gain entry into markets with higher standards. Freeing up trade can improve environmental standards, too. In contrast, trade barriers, from tariffs to embargoes, can serve to lower them, shielding incompetence by distorting market signals. Firms manufacturing behind trade barriers face less competition, often with fewer incentives to upgrade facilities or avoid unnecessary waste. Government subsidies, such as tax relief for farmers growing feed grain, can also cause financial and environmental inefficiencies.

International financing organizations can directly support poorer states striving to implement environmental policies. One example is the Global Environment Facility (GEF), which began as a pilot program in 1991, and which now involves three implementing agencies: the World Bank, the United Nations Development Programme, and the United Nations Environment Programme. Its mandate is to disburse grants and technical assistance to developing countries for projects having a global environmental goal (such as mitigating climate change or protecting biodiversity). As one of the world's major sources of financing for such projects, the GEF has distributed more than $7 billion in grants and $28 billion in cofinancing from other sources.

Thus a globalizing world economy, coupled with globalized environmental policies and institutions, can improve environmental management—and is already doing so on some measures. As the next section explains, however, this "progress" relies, at least in part, on an unbalanced process of economic globalization that draws down natural resources and deflects the costs of rising consumption away from those who benefit the most and toward those who benefit the least. Therefore environmental progress may appear to occur in one location (London, Paris, Los Angeles), while another location (New Delhi, Rio de Janeiro, the open oceans) absorbs the resulting costs. This helps to explain why the *total* stress on the biosphere is rising even as the life-cycle impact of *particular* consumer goods declines.

Unbalanced Globalization

Globalization is widening what some call the "global market" and others the "global consumer culture." A few statistics will suffice to show this. The value of world merchandise exports now exceeds $10 trillion, up from $6 trillion in 2000, an amount that even then was over 100 times higher than in 1948. On average, foreign currency trading is now around $2 trillion per day, up from about $1 trillion per day a decade ago, and far higher than the daily trading of $10–20 billion during the 1970s. The number of multinational parent companies with investments in more than one country has grown, too, from about 7,000 in 1970 to over 78,000 today (with more than 780,000 affiliate firms). The flow of foreign direct investment into developing countries has been rising steadily as these multinationals continue to expand: from $22 billion in 1990 to $380 billion in 2006 (the highest ever).[11]

Trade and multinational corporations have been engines of growth for the world economy. Figures over the last few decades show the rapid rate of economic growth during a time of increasing trade and growing numbers of multinational corporations. World GDP (in constant 1995 dollars) almost tripled from 1970 to 2000: from $13.4 trillion to $34.1 trillion. The world economy continues to expand, too. It grew more from 2001 to 2006 than in any five-year period since World War II. Over this time, First World economies grew on average by over 3 percent. Growth in the Third World was even faster, with an average expansion of about 7 percent in 2006 (following 6.6 percent in 2005 and 7.2 percent in 2004).

As globalization intensifies, national incomes could grow even faster in the next 25 years than in 1980–2005. In what it describes as the "next wave of globalization," the World Bank predicts the output of the global economy—led by growth in developing countries like China and India— could well expand from $35 trillion in 2005 to $72 trillion in 2030 (in constant exchange rates and prices). This assumes an annual average growth of 2.5 percent in developed countries and 4.2 percent in developing countries. The World Bank also expects a more than threefold rise in global trade in goods and services by 2030 (to $27 trillion). Over this period, it expects trade as a share of the global economy to jump from one-quarter to more than one-third. Such rapid economic growth within developing countries alone would increase the number of "middle-class" consumers, with incomes of between $4,000 and $17,000, from 400 million to 1.2 billion. Maintaining such growth over the next 25 years would increase the purchasing power of billions of people, allowing the new middle class, for example, to afford advanced consumer goods like automobiles. By then, assuming trends hold, the World Bank predicts average living standards in countries like China, Mexico, and Turkey will more or less reach those in Spain today. The number of people living in "dire" poverty—defined as having an income of less than $1 a day— would also fall from about 1.1 billion to 550 million (despite growing populations in the poorest countries).[12]

So far, the "flattening" of the globe under globalization has not, however, brought equal or balanced outcomes for individuals or societies.[13] The superrich like Bill Gates now live on an island in a sea of 2.7 billion people subsisting on less than $2 per day. Over 800 million people, in a world with 946 billionaires worth $3.5 trillion in 2006, continue to suffer from chronic malnutrition. Over a billion people do not even have access to clean water. *Forbes* magazine ranked Americans Bill Gates and Warren Buffett as the world's richest in 2006. Together, these men were worth $108 billion ($56 and $52 billion, respectively). The world's third richest was not far behind at $49 billion. Tellingly for the unequal effects of globalization, this was Carlos Slim Helú, a citizen and resident of Mexico. The forces generating much of this unequal wealth—corporations, trade, and financing—are, as the next three sections show, especially prone to deflecting ecological costs of consumption away from the wealthy and toward the poor and powerless, a process that may partly explain why some people try to resist, or on occasion fight to reverse, globalization.

Side Effects of Corporate Behavior

Since at least colonial times, entrepreneurs have been heading overseas to supply consumers back home. The risks have been considerable, but so have the profits from cheap or exotic products. The first wave of entrepreneurs from Europe in the 1600s went to "collect" natural resources or "harvest" crops. Over the next few centuries, loggers from Britain trekked into the rainforests of Southeast Asia for teak and mahogany, miners from France dug deep into the heart of Africa for diamonds and gold, and fishermen from these and other European nations sailed across the Atlantic for cod and seals. Overseas plantations also began to supply wealthy consumers in Europe with luxuries like tea, coffee, bananas, sugar, and pepper—luxuries that soon became "necessities." Over time, manufacturers also began to move overseas to gain access to these inexpensive natural resources, as well as to cheap labor and infrastructure. Imports of natural resources and goods into Europe increased markedly in the late eighteenth century. By the turn of the twentieth century, global imports were accelerating even faster, as the United States began to surpass the import might of the European economies.[14]

The history of trade and financing was marked by racism, brutal wars, and cultural annihilation. Although the multinational companies and trade chains emerging from this period have benefited some developing countries in some ways, many of the structural imbalances of relations between them remain, and, if anything, the intensity and range of the ecological shadows arising from these corporate activities is even greater today than during colonial times. One reason is the sheer number of multinational corporations today, including ones increasingly from developing countries like Malaysia (for example, in the logging industry) and China (for example, in the mining industry). Already, by the beginning of the twenty-first century, multinational corporations accounted for one-tenth of world GDP, while intra-firm trade accounted for one-third of world exports.[15] Since then, with economic globalization opening markets and encouraging mergers and acquisitions, the financial clout of the biggest corporations—Citigroup, General Electric, Exxon Mobil, Wal-Mart, Microsoft, Ford, General Motors—has continued to grow.

When investing, these companies tend to bring along "advanced" technologies, financing, and access to global markets. Such investment tends to promote "efficiencies" and expand markets, and thus to grow economies. But, in doing so, these multinationals also tend to extract more natural resources, whether timber, fish, or minerals, or to manufacture more goods than local firms; many do so as quickly as possible, exporting to wealthy markets to earn foreign exchange. Many agricultural companies continue to invest in the production of food for export, in palm oil plantations on the outer islands of Indonesia (to supply the margarine and fast-food cooking oil markets), for example, or in cattle ranches in the Amazon rainforests (to supply the growing global demand for cheap beef). Such operations commonly rely on chemicals and fertilizers—inputs most often supplied by distant multinationals—with long-term environmental costs.

Shifting toward export crops and a reliance on chemicals and fertilizers can cause nutritional deficiencies among local people as subsistence farming declines. Moreover, mass-producing goods in developing countries with low environmental standards tends to pollute rivers, soils, and water supplies. Some corporations in the First World are also shipping garbage and hazardous waste to poorer countries (such as computer waste to China). Everywhere, corporations also spew dangerous substances (dioxins, furans, PCBs, DDT) into the air, poisons that eventually fall back to earth. Granted, virtually every multinational corporation now works within an environmental plan, labeling it something like "sustainable yield" or "sustainable management" or "sustainable investment." Yet, in many cases, these plans are either unrealistic or put in place largely, if not entirely, for public relations. Although, over time, competition among multinational companies, even in developing countries, does tend to produce less environmental harm per unit of output, at the same time, it tends to expand markets, which in turn casts larger ecological shadows of consumption even as per unit damage declines.

Multinational corporations cast ecological shadows in other ways as well. They tend to work within complex trade chains of suppliers, financers, producers, wholesalers, and retailers. Such networks tend to reduce accountability and transparency, making it exceedingly difficult to hold any single corporate entity accountable for environmental costs. Some of these firms hide costs through illegal activities. Among multinational companies logging the rainforests of Southeast Asia and the South Pacific, for example, smuggling, evading taxes, and transfer pricing are all commonplace, as are bribes to enforcement officers, customs officials, military officers, and politicians.

Almost all multinational companies also employ double standards, obeying a higher set at home than in host countries. This is particularly true for labor standards, such as wages, pensions, and accident insurance, but it's also the case for many environmental standards. Although, arguably, double standards allow companies to "respect" local laws and traditions, countenancing these double standards means that many national policies designed to protect citizens from environmental harm serve, instead, as incentives for companies to expand into overseas markets with lower standards. Governments with the higher standards commonly ignore such "unintended consequences" to appease corporate opposition and allow for more effective domestic implementation of environmental regulations, while ensuring that companies remain profitable. For these reasons, the double standards of multinational corporations remain a core force in casting ecological shadows of consumption onto the poorest regions of the world.

The aggressive pursuit of corporate profits tends to cast even longer ecological shadows. State incentives to sustain economic growth—or the lack of penalties for actions harmful to the environment—can reinforce such tendencies. So, too, can the efforts of advertisers to expand or hold markets through strategies like branding. Firms may profit from intentionally deflecting environmental costs, for example, by dumping untreated waste into nearby rivers. Firms can profit as well from introducing new goods or services with uncertain risks, in effect, experimenting with the health of consumers or the integrity of ecosystems. Some of the most profitable corporate "innovations" of the last century, such as the discovery in 1928 of "safe" and "stable" chlorofluorocarbons (CFCs) for refrigerators and air conditioners, have dispersed lasting harm far into the future. In this case, the consequences for depleting the ozone layer were unknown to science for over four decades.

But, for countless other innovations, such as adding tetraethyl lead to gasoline after the 1920s, the possible consequences alarmed some scientists right from the beginning. In such cases, industry scientists and corporate spinmasters work hard to keep critics on the defensive. Corporate executives lobby politicians, demand "proof" of *direct* harm, raise "scientific" doubt, and rely on drawn-out legal battles to delay regulations or phaseouts. It's during these "tough" times at home that some multinational corporations expand their overseas markets (as in the case of the tobacco industry). Firms may also begin to search for profitable substitutes. Sometimes, these substitutes eliminate the harm; but other times, they begin the process of experimenting on consumers all over again with yet another "improved" product (such as benzene in unleaded gasoline or hydrofluorocarbons in refrigerators). Corporate research teams are always searching for the next cutting-edge process or new product able to capture or expand a market. The effects of introducing many of these "innovations" may remain uncertain for years, sometimes generations. Who, for example, can really predict the future consequences of genetically modified organisms or nanotechnology?

Side Effects of Trade

The globalization of trade can interact with corporations to lengthen ecological shadows in other ways as well. Producer and consumer prices of many traded products, such as for timber and beef, do not reflect the full ecological or social costs of harvesting, processing, producing, marketing, or disposal. The resulting low consumer prices can in turn contribute to wasteful consumption and overconsumption (defined as consumption with no benefits for well-being, such as overeating until obese). This helps to explain why strategies like "supersizing" are so profitable for fast-food chains, and why rates of obesity are rising world-wide, even as millions of people continue to starve.

Wasteful consumption in turn feeds back into more expansive and damaging ecological shadows. The environmental history of Japan's rapid growth after World War II illustrates this well. Japanese trading companies—such as the Mitsubishi and Sumitomo Corporations—began financing networks of firms to import large quantities of cheap natural resources into the fast-growing Japanese economy. Much of these resources, as in the case of wood, came from Southeast Asia. Japan imported, for example, 60 percent of total log production during the height of the logging booms in the old-growth forests of the Philippines (1964–1973) and the Malaysian state of Sabah (1972–1987), and 40 percent during the boom in log exports from Indonesia (1970–1980). These firms turned next to the Malaysian state of Sarawak and to Mela-nesia after cheap and accessible log supplies declined sharply in the Philippines and Sabah, and after Indonesia restricted raw log exports to prop up its domestic plywood industry. By the mid-1990s, Japan was absorbing about half of the total log exports from Sarawak, Papua New Guinea, and the Solomon Islands. Japanese processors turned the bulk of these raw logs into plywood panels for a booming construction indus-try looking for inexpensive ways to mold concrete. These panels, known as "kon pane" in Japanese, were generally burned or left to rot after only a few uses. The reason for such "waste" was straightforward: it was cheaper to buy new panels than clean the old ones.

Japan had no Machiavellian plot to protect its own forests at the expense of others'. Japanese firms went searching overseas for timber because supplies within Japan were inadequate (insufficient or of lower quality) and more expensive. Also, Japanese consumption is not solely to blame: deforestation, biodiversity loss, and soil erosion that continue to this day to sweep across Southeast Asia and Melanesia are a result of a complex interplay of political, socioeconomic, and ecological forces from the global to local levels. Still, the boast in the early 1990s by the Japanese Forestry Agency that Japan was now one of "the most heavily forested countries in the world" reveals much about the delusional accounting of environmental progress still so common today among national governments.[16]

Global trade can lengthen ecological shadows in other ways, too. Governments in an effort to increase trade or participate in trade agreements will sometimes lower—or perhaps fail to strengthen—environmental rules. This can cause a "race to the bottom" among states; or, in some cases, leave countries "stuck at the bottom."[17] Moreover, what wealthy states label as "sound trade practices"—even after the so-called liberalization of trade—is hardly "free" or "fair," with many trade rules continuing to protect the interests of powerful countries (such as farmers in the United States and western Europe). The globalization of trade is also lengthening the distances between producers and consumers, so that users don't perceive—or at least can more easily ignore—the effects.[18] It's creating larger and more diverse markets as well, casting shadows into increasingly distant lands. Even if markets collapse, as happened after the European ban on the import of whitecoat harp seal pelts in the 1980s, new markets can readily form with even more consumer demand, as is now happening in China and Russia for seal furs.

Side Effects of Financing

Foreign aid buttresses many of the trading and corporate structures casting ecological shadows into developing countries. Organizations like the World Bank and IMF impose conditions on assistance requiring governments to liberalize trade and investment. No doubt reforms such as eliminating tariff barriers can improve environmental management by reducing waste and inefficiencies. Yet decades of foreign aid have left much of the developing world with ballooning foreign debts. Total external debt for developing countries stood at just over $72 billion in 1970. Ten years later, it was over $600 billion. A decade after this, it was nearly $1.5 trillion. By the beginning of the twenty-first century, it was hovering around $2.4 trillion. Total debt service paid by developing countries in 2001 was more than $377 billion, of which $116 billion was in interest repayments. Since then, even with recent international efforts to provide some relief for the most heavily indebted countries, debt burdens in more than half of developing countries have continued to worsen.[19]

Such heavy foreign debts push governments to pursue development paths that allow them to earn sufficient foreign exchange to service or repay loans. This tends to mean strategies like exporting gold, timber, and oil, or clearing land for cash crops like coffee and sugar. It tends as well to encourage governments in developing countries to build infrastructure to entice further investment in natural resources, plantations, and low-end manufacturing. All of this yet again reinforces a globalizing order that deflects the ecological costs of consumption to distant places and times.

Deflecting Costs and Responsibilities

Within countries, these costs tend to be deflected into places like industrial neighborhoods or indigenous communities: an outcome often reinforcing existing patterns of inequality and racism. Across nations, they tend to be shifted from wealthy states and cities to poorer countries and regions: an outcome aggravating existing South-North inequities. More globally, costs tend to gravitate toward places far from centers of power: into the deserts of Africa, the rainforests of the Amazon, the sea life of the Arctic, the depths of the Pacific Ocean, the heights of the stratosphere. Costs tend, as well, to drift into the future, often with far greater consequences: accumulating in ecosystems and exposing people to long-term health risks (with the poor facing higher risks). All of these changes can reinforce patterns of wasteful and excessive consumption. At the same time, shifting environmental consequences into poor communities can undermine social well-being by contributing to inadequate nutrition or insufficient housing and can cause political and economic instabilities, creating the potential for weak states or fragile environments to spiral into collapse.

Governing the consequences of these ecological shadows of consumption is much harder than stopping a chemical plant from poisoning a stream or citizens from tossing garbage into the streets. Often, the processes channeling these consequences occur inside a global system so complex, so chaotic, that tracing the pathways of cause and effect is beyond the traditional tools of policy makers or scientists. The domino effects of resulting changes can also create many unpredictable outcomes, with consequences snowballing within and across various systems. This tends as well to disperse and distance responsibility, leaving many consumers unable to perceive the differences for the global environment of choosing among various options. Who, after all, is really accountable or responsible for the collapse of the 120-foot-thick Ayles ice shelf?

Of course, to some extent every consumer is responsible, although not all share equal responsibility. Those with power and wealth are consuming far more of the world's ecological resources: a life of luxury in Philadelphia deflects more environmental damage farther than a life of poverty in Harare. Still, no single consumer, no matter how wasteful or profligate, can cause an ecological shadow to form or shift direction, although this does not absolve consumers who ignore the effects of their personal choices on the sustainability of life for others. Accepting that these effects are "real" is essential for sustaining the collective will for reforms.

Yet far-reaching change will require far more than educating some consumers in some cultures to consume a few things more thoughtfully. As this chapter reveals, it will require tackling structural features of a world order that deflects environmental costs of consumption into spaces with relatively less power. In particular, governing mechanisms will need to guide globalization more effectively, strengthening environmentalism in ways that rein in the shadow effects of corporations, trade, financing, and local policies. Immediate action is imperative. As the global population races toward 9–11 billion, worldwide economic growth shows every sign of racing even faster, global consumerism every sign of consolidating further, and the next wave of globalization every sign of increasing both the scale and the speed of the ecological changes brought about by the shifting global patterns of consumption.

In all likelihood, the globalization of environmentalism will continue both to improve the efficiency of producing, using, and recycling consumer goods and to promote further advances in global governance, from greener corporate codes of conduct, to stricter international environmental laws, to stronger cultural norms of "appropriate" consumption. But, as things now stand, and as chapter 23 will elaborate in the conclusion to this book, it will do so at a pace that is too slow and too incremental to prevent the intensity and spread of ecological shadows from escalating. The costs to the global environment and human health, as chapter 2 will make clear, are already too great *not* to take immediate action.

2

Dying of Consumption

The unequal globalization of the costs of consumption is putting ecosystems and billions of people at risk. Many living within small worlds of prosperity, however, end up seeing more progress than peril around them, pointing to better environmental practices and technologies, to energy-efficient appliances, greener architecture, organic foods. Relatively few in power ever question the side effects of a global political economy producing ever more "new and improved" products—even as threats to just about every ecosystem continue to escalate.

The Darkening Skies

Many natural environments are in crisis.[1] Over half of the world's original forests and wetlands are now gone. The tropical rainforests, wonders of biodiversity, remain under severe threat from loggers and industrial farmers. The tropics are now losing over 13 million hectares (32 million acres) of natural forest every year: Brazil alone is losing over 3 million hectares (7 million acres), while Indonesia is losing nearly 2 million (5 million acres).[2] Meanwhile, the once seemingly infinite oceans now swirl with toxins like mercury. Such pollution has done little to discourage fishing, which persists at levels that are pushing many commercial stocks into collapse. The number of Atlantic bluefin tuna, for example, has fallen by at least 80 percent since 1970.[3] The northern cod off the eastern coast of Canada—whose numbers were so plentiful in the 1600s that sailors could fill a bucket simply by lowering it over the side—is now endangered, its population falling by 99 percent over the last four decades.[4]

Such tales are becoming common across the globe. In its analysis of data on 7,800 species of wild seafood, a 14-person team found worldwide catches of 29 percent of these species are now at least 90 percent below past averages. Unless measures are taken to curtail harvests, the team predicts a "collapse" of much of the remaining commercial wild seafood before 2050.[5] Another 10-year survey of the global oceans found a 90 percent decline in large predatory fish over the last half century, including cod, flounder, marlin, swordfish, and tuna.[6] Indeed, unless the world changes course, the waters of Ernest Hemingway's *Old Man and the Sea* will soon be empty of the fighting marlin, an outcome arising, not from the heroics of men like Hemingway's old man Santiago, but from industrial fishing boats plying the oceans to feed global markets.

Land and freshwater resources are also under great ecological pressures from human activities. The fate of the Aral Sea illustrates the magnitude of some of these changes: once the world's fourth largest lake, it shrank by half in just three decades as its inflows were diverted to agriculture and hydroelectricity, leaving it as salty as an ocean by the early 1990s. Demand from agriculture, industry, and individuals is continuing to deplete scarce water resources elsewhere, too. Over a billion people are now struggling to survive without access to clean water, and current trends suggest billions more will live with severe water shortages within a few decades.

Changes like these are contributing to the death of between 50 and 150 species every day. Many are microscopic, deep in oceans and forests, and still beyond the reach of scientific cataloguers trying to deal with the estimated 5–30 million species of life on earth. Yet even macroscopic plant and animal species are going extinct, at an average rate 50–100 times higher than the natural one (assuming an average life span of 5–10 million years for a species)—over 1,000 since the beginning of the seventeenth century. The Worldwatch Institute believes the planet is now "in the midst of the biggest wave of animal extinctions since the dinosaurs disappeared 65 million years ago."[7]

The exponential growth in consumption is also saturating the global environment with chemicals. Some, like DDT to kill mosquitoes that spread malaria, save millions of lives every year. Yet these "useful" chemicals are also contaminating ecosystems and poisoning people. Some 75,000 chemicals are registered in the United States alone, yet, of these, scientists have tested the carcinogenicity of only 1,500 (or 2 percent). They know even less about the toxicity of the 11,000 commercial organochlorines or the thousands of accidental—and often unknown—chemical by-products.[8]

One of the greatest sources of chemicals is agriculture. Over the last half century, farmers have come to rely more and more on pesticides and fertilizers. American farmers, for example, were using 50 million pounds of pesticides per year in the 1940s. By the end of the 1970s, it was over 800 million pounds. Globally, from 1961 to 1999, the use of pesticides went up more than 800 percent. In the same period, the use of nitrogenous fertilizers went up more than 600 percent; that of phosphate fertilizers, more than 200 percent.[9] Such growth in the use of chemicals has most likely played a part in the rising rates of diseases like cancer, but few governments or firms seem eager to investigate the environmental sources of such diseases, focusing instead on diagnosis, treatment, and cures.

Climate change is perhaps the greatest environmental threat of all.[10] In a 700-page report commissioned by Britain's chancellor of the exchequer, former chief economist of the World Bank Nicholas Stern put the potential economic and social disruption of global climate change on a par with that of both World Wars and the Great Depression combined.[11] Just about every aspect of modern consumer life—manufacturing, traveling, heating, cooling, burning, eating—is producing greenhouse gases, notably carbon dioxide, methane, and nitrous oxide. Deforestation is releasing carbon dioxide, too, and now accounts for 25 percent of anthropogenic emissions of carbon dioxide. As a result of all of these activities, total carbon dioxide emissions increased twelvefold over the twentieth century.[12] In the twenty-first century, the rate of increase for carbon dioxide emissions from burning fossil fuels and making cement has more than doubled, from an average of 1.3 percent per year in the 1990s to 3.3 percent per year from 2000 to 2006. The jump in carbon dioxide emissions from 2000 to 2006, according to a study published in the *Proceedings of the National Academy of Sciences*, was the fastest rate of increase over a seven-year period since modern records began at the end of the 1950s. The concentration of carbon dioxide in the atmosphere is now over 380 parts per million—the highest level in at least 650,000 years (and perhaps the highest in 20 million years).[13]

Greenhouse gases, with global emission rates now over 70 percent higher than in 1970, are warming the planet.[14] The earth's average surface temperature rose by about 0.6 degrees Celsius (1.1 degrees Fahrenheit) over the twentieth century. This may not seem like much. Yet it made the twentieth century the warmest one of the last millennium. One obvious sign of global warming is the melting polar ice caps, which have been shrinking by about 9 percent every decade since 1979. Another is the recent melting of the 11,000-year-old permafrost in western Siberia. And the problem of rising temperatures seems to be worsening. The 1990s was the warmest decade and 2005 the warmest year over at least the last century. Records were broken in just about every year of the last decade. Tied as the second and third warmest years of all time were 2007 and 1998 (a year with a strong El Niño); the fourth warmest was 2002 (a year with a weak El Niño), followed by 2003 and 2006.[15]

The twenty-first century will likely be even warmer. Six scenarios by the United Nations Intergovernmental Panel on Climate Change (released in 2007) show a likely rise in the average worldwide surface temperature over the next century of another 1.1–6.4 degrees Celsius (2.0–11.5 degrees Fahrenheit) from the 1980–1999 average—with best estimates pointing to the fastest rate of change for at least the last 10,000 years. A rise of 3–5 degrees Celsius (5.5–9 degrees Fahrenheit) would, according to NASA's Drew Shindell, "bring us up to the warmest temperatures the world has experienced probably in the last million years."[16]

The future may see even warmer temperatures, however, if the process reaches a "tipping point."[17] Some scientists now worry that global warming is diminishing the capacity of the earth's "sinks" (land, forests, and oceans) to absorb or retain greenhouse gases. Two examples of the latter are in the Antarctic Ocean, where stronger winds linked to warmer temperatures are now churning up waters rich in carbon dioxide, and in Siberia, where the melting permafrost is releasing methane, a gas with 20 times the greenhouse effect of carbon dioxide. A warmer world means more of this permafrost will melt, which will release more methane, which will raise temperatures, which will melt more permafrost. This self-reinforcing feedback could release around 49 billion metric tons of methane (nearly one-sixth of all of the world's methane stored on land) from the northeast Siberian ice complex alone.[18] Other self-reinforcing feedbacks could further accelerate warming.

Warmer temperatures will have many unpredictable and uneven consequences. Wind, rain, and snow patterns will change, with some places becoming hotter and some colder. Rising oceans will engulf low-lying islands. Droughts will disrupt agricultural yields, especially in places like Africa. Severe weather—hurricanes, tornados, hailstorms, droughts— will occur more frequently and with greater intensity. Although the world is unlikely to see the next ice age charge down the streets of New York City like a giant grizzly bear, as it does in the 2004 movie *The Day After Tomorrow*, global warming will be catastrophic for many species. A temperature rise of just 0.8–2.0 degrees Celsius (1.4–3.6 degrees Fahrenheit), for example, could "commit" 18–35 percent of plant and animal species to extinction by 2050. Other factors, like higher concentrations of carbon dioxide, could lead to even higher rates of extinction.[19]

Changes to the global environment are already harming billions of people, from the Inuit in the Arctic to the Penan of Sarawak to the Brazilians of Rio de Janeiro. One example is the more than 10 million children under the age of five who are dying every year from preventable and treatable causes, with unhealthy environments contributing to almost half of these deaths.

Just as disturbing, many of us are being exposed to health risks as firms experiment on consumers with a rush of new products. The international legal community rightly applauds its success in phasing out chemicals like chlorofluorocarbons (CFCs). But what about the thousands of other chemical "discoveries" now in our food, air, and water? What will happen when these chemicals combine? Some of the chemicals will prove harmless. But some will prove harmful and some even deadly. Scientists are testing, arguing, and analyzing, as are firms, activists, and government agencies. As with CFCs, it will take years, perhaps decades, to see the full consequences of introducing these chemicals into our environments.

Examples of substances with the potential to harm ecosystems—and thus human health—seem to trickle into the daily press in a steady flow. Some are the result of an activist group or reporter sensationalizing a story. But many arise from scientific tests producing truly worrisome results. As the next section shows, current debates over the use of PBDEs in furnishings and electronic devices call to mind those of decades past over the use of DDT around the home to keep mosquitoes at bay.

Consuming Risks

The DDT, PCBs, and CFCs of today include chemicals like PBDEs (polybrominated diphenyl ethers). For over three decades, firms have put PBDEs into household and office items (mattresses, pillows, rugs, curtains, carpet padding, TVs, computers) as flame retardants. These chemicals were heralded as a great advance in consumer safety in the 1970s, able to prevent a TV from bursting into flame or slow a fire in a mattress. Back then, chemists and medical specialists could see few reasons to worry about putting PBDEs inside hard plastics or soft foams. These were only toxic in large quantities; besides, there was no reason to expect they wouldn't remain safely inside a product.

Swedish scientists set off alarms in the late 1990s after discovering that concentrations of PBDEs in human breast milk were rising in some populations. Soon it was clear these chemicals were migrating from consumer products into humans—not, it seems, primarily through food chains as with other persistent organic pollutants like dioxins or PCBs, but by collecting in home environments, especially in indoor air and household dust.[20] Recent tests show U.S. residents now have the highest levels of PBDEs in the world (followed by Canadians). Residents of North America, on average, have 10–70 times higher PBDE levels than residents of Japan or Europe. Some individuals—between 5 and 10 percent of the North American population—appear to have absorbed especially large quantities, perhaps because of exposure to crumbling foam in mattresses and furniture, or perhaps because of exposure to dust as crawling babies. Tests of breast milk, tissue, and blood show these individuals have levels of PBDEs around 1,000 times higher than those with low readings.

In laboratory experiments, such high levels cause symptoms in animals similar to those of hyperactivity and attention deficit in children. PBDEs appear to lower sperm counts as well. Although their chemical structure resembles that of PCBs (some would call them "chemical cousins"), medical researchers are focusing on the unique qualities of PBDEs, which appear to mimic and interfere with human hormones (such as thyroid hormones). Some specialists now think that, unlike typical toxic chemicals, PBDEs may damage the brain in only trace amounts—which doctors would have thought inconsequential in the past—provided exposure occurs at a critical juncture in growth. Recent experiments to test the effect of trace amounts of PBDEs—amounts already present in some humans—found permanent brain damage in rats and mice.

Over the last decade, European governments have taken steps to eliminate two particularly worrisome formulations of PBDEs: those commonly used in mattresses, on the one hand, and in computer housings and monitors (representing about 15 percent of the global market for PBDEs in 2001), on the other. A number of states in the United States have done the same. The Environmental Protection Agency (EPA) has managed to encourage some of the major producers of PBDEs to phase out these two formulations voluntarily. Some manufacturing and retail firms are also taking steps to stop using PBDEs. The Swedish home furnishing company IKEA was one of the first firms to remove them from its products. The U.S. computer company Dell and the Swedish automaker Volvo are examples of firms now working toward eliminating some of the worst formulations of PBDEs.[21]

The latest research findings on the role of pesticides and herbicides in neurodegenerative illnesses are just as alarming as those on PBDEs. In a 2006 epidemiological analysis of the 143,000 participants in an ongoing study by the American Cancer Society, for example, researchers found those regularly exposed to low doses of pesticides and herbicides—such as gardeners, farmers, ranchers, and fishers—had a 70 percent higher incidence of Parkinson's disease than those not so exposed.[22]

Perfluorochemicals are another family of chemicals with worrisome properties for the health of consumers. Researchers are focusing on two members of this family: perfluorooctanoic acid (PFOA) and perfluorooctanyl sulfonate (PFOS). Virtually indestructible, they are used to make Teflon pots and pans nonstick and to make rugs, couches, and raincoats grease-resistant, stain-resistant, and waterproof. You can also find them in pizza boxes, microwave popcorn bags, fast-food burger wrappers, and French-fry containers, as well as in nail polishes and shaving creams.

PFOA and PFOS are migrating (exactly how is still unclear) from consumer goods into the environment and into humans, where, like lead, PBDEs, and pesticides, they are appearing in detectable quantities. Again, as with these other chemicals, an increasing number of doctors now see exposure to low doses over long periods as a potential health threat, especially for children. Recent laboratory tests of PFOA on animals, for example, have found links to low birth weights, damage to thyroid glands, changes in male reproductive hormones, breast cancer, and liver cancer. As tests in the United States and other countries find PFOA and PFOS in some children at levels above those causing measurable harm in laboratory animals, more and more health and environmental specialists are becoming alarmed.[23]

Such findings are particularly disturbing because natural biological processes do not appear to ever break these chemicals down into less harmful substances, as they do with chemicals like DDT. Richard Wiles, vice president of the nongovernmental Environmental Working Group, calls them "the DDT of this millennium," but with much higher stakes because they "last forever." Although humans appear able to excrete PFOA over a period of several decades, still, the wonder chemical that nothing can stick to, seems able to stick to living things long enough for tiny quantities to bioaccumulate until toxic.

DuPont insists its Teflon pots and frying pans are safe if consumers use them properly. Manufacturers use PFOA to produce Teflon, corporate brochures explain, but it's not an "ingredient" in the Teflon itself. David Boothe, DuPont's global manager for products like Teflon, explains: "When you're using the cookware as it's intended to be used, at the temperatures it's intended to be used, it's perfectly safe."

Many companies—even DuPont—are nevertheless beginning to give up on these chemicals. The first to do so was the U.S. technology firm Minnesota Mining and Manufacturing (called 3M since 2002), when in May 2000, after four decades of production, it began phasing out PFOS from the popular brand Scotchgard. It took steps as well to stop producing PFOA. On the same day 3M made these announcements, the EPA informed governments worldwide of animal tests showing that PFOS "appears to combine persistence, bioaccumulation and toxicity properties to an extraordinary degree." Since 2000, a few other firms have been following 3M's lead in voluntarily reducing use, such as the fast-food chain McDonald's, which no longer uses wrappers containing perfluorochemicals.[24]

Some governments are also now beginning to push firms harder to get rid of PFOA. Both Canada and the United States, for example, began initiating phasedowns and safety reviews of PFOA in 2006. The EPA reached a deal in early 2006 with DuPont and seven other manufacturers to reduce PFOA emissions from their U.S. facilities and PFOA in their products by 95 percent by 2010, with the goal of completely eliminating them by 2015. DuPont took swift measures, cutting back PFOA production by 95 percent in 2006.

Safety reviews of hundreds of other chemicals in widespread use are also now occurring across much of the developed world. On that list is bisphenol A (BPA), a synthetic petrochemical and a main ingredient in polycarbonate plastics and resins. Volumes have been growing steadily since the 1950s—with major producers including Bayer, Dow Chemical, GE Plastics, and Sunoco—and it's now one of the world's most common chemicals in production. Hard, clear plastic water and baby bottles contain bisphenol A. So do compact disks, sports helmets, microwavable plastics, dental sealants, and the lining of many tin cans. Like so many other chemicals, trace amounts of bisphenol A are migrating into the environment and into people. Scientists have known since the 1930s that it can mimic the female hormone estrogen. Still, the scientific consensus for much of the twentieth century was solid: the level of exposure was low and thus posed no danger to health.

Hundreds of experiments over the last decade, however, have found possible links between bisphenol A and prostate cancer, diabetes, low sperm counts, and the early onset of puberty. The converging findings of much of this research challenge a commonsense tenet of toxicology going back to at least the fifteenth century: that a higher dose equals *more* harm. An increasing number of scientists are coming to conclude that trace amounts of bisphenol A, which the body treats as a hormone (turning on receptors in cells), may in fact cause more harm than larger quantities, which the body treats as a poison (causing receptors to overload and stop functioning). Class action lawsuits are just beginning to draw on this research. The first was against five manufacturers of baby bottles, filed in Los Angeles in March 2007. Cases like this one will raise further questions about the safety of numerous other consumer products containing chemicals that can disrupt hormonal systems.

Efforts to phase out chemicals like PBDEs, PFOA, and BPA will no doubt take considerable time. Manufacturers like DuPont will continue to insist on "science-based approaches" (including funding research). Litigation will drag on, as firms file appeals and countersuits. And the progress across different jurisdictions will inevitably be uneven as these same firms shuffle risk overseas to keep sales and profits healthy. Steven Hentges of the American Plastics Council sounds no different today than most of the other corporate spokespeople of the last century: "BPA is not a risk to human health at the extremely low levels at which people might be exposed from use of, for example, polycarbonate plastic." The reaction of scientists sounds equally familiar. Biology professor Frederick vom Saal, a specialist on hormones and synthetic chemicals, responds: "The chemical companies think they can lie with impunity about the published scientific literature."[25]

Meanwhile, as these debates rage, other chemicals with other side effects, by themselves and in combinations, are entering the global marketplace, adding further to the total ecological burden, with unpredictable consequences for human and environmental health. Clearly, particular chemicals and combinations of chemicals are harming the health of some people, although the pathways of causality are so complex it's impossible to determine precisely how and to what extent. Still, a glance at global cancer rates reveals some disquieting trends.

A Shadow of Cancer?

Could the annual worldwide use of 400 million metric tons of chemicals in part explain the rising cancer rates? What about other environmental changes like ozone depletion, air and water pollution, or climate change? Or factors like the increasing consumption of processed and fast food? Globally, around eight million die every year from cancer—a figure the

World Health Organization predicts will increase to more than 10 million over the next two decades (with more than 15 million new patients a year). It remains, in the prophetic words of French surgeon Stanislas Tanchou in 1843, a "disease of civilization."

Cancer is now the second most common cause of death in the developed world, after cardiovascular disease. Cancer rates in the United States, even after adjusting for longer life expectancies and excluding lung cancer, have been rising steadily (several studies put the increase at around 35 percent since the 1950s).[26] Cancer is now responsible for almost one-quarter of all deaths in the United States. The American Cancer Society estimates the lifetime chance of getting cancer in the United States at nearly one in two for men and just over one in three for women.

Why are cancer rates rising? The reasons are complex: eating habits, exercise choices, ever better diagnostic techniques. It's certainly simplistic to blame chemicals or power lines or pollution alone. Still, it seems sensible to worry about the thousands of recent laboratory results linking cancer in animals to chemicals common in consumer products (from shampoo to gasoline to French fries), drinking water, and air (both indoor and outdoor).[27] I find it just as sensible—even without scientific proof of causality—to worry about the brews of environmental toxins scientists are now finding in the bloodstreams of many people.

How can consumers avoid exposure to chemicals with potentially deadly effects? There's only one way, Ana Soto of Tufts University School of Medicine wryly explains: "Don't eat, don't drink, and don't breathe."[28]

Mapping Ecological Shadows

At the start of the twentieth century, average life expectancy was 30. Today, it's over 66, and just listing all of the medical advances of the last 100 years—antibiotics, obstetrics, heart transplants, pasteurization, vaccines—could fill a book. Still, this does not excuse governments and companies for failing to do more to protect consumers and ecosystems. But to understand what action must be taken, we need to map particular shadows of consumption in detail—to learn how they are affecting us and *why* they are advancing or receding.

Chapters 3–22 do this by analyzing the evolution of how automobiles, gasoline, refrigerators, beef, and seals have been made, raised, or hunted and how they have been consumed. The chapters cover diverse geographies, eras, sectors, governance structures, and political economies (from low-end natural resource extraction in the seventeenth century to high-end manufacturing in the twenty-first century). Without doubt, however, even this wide-ranging set of cases does not exhaust the nuances of change occurring in the thousands of political economies of consumption. Analyzing other consumer goods, from coffee, bananas, sugar, and tea to whales, elephants, tigers, and pigs would help shed light on some of those nuances. So would analyzing sectors like fisheries, forestry, and mining or issues like biodiversity, pesticides, coal, hazardous waste, and persistent organic pollutants.[29] Still, the five cases that follow survey enough ground in enough depth to reveal the consistent forces of environmental change and the consequences of ecological shadows.

In every case, environmental management is improving some matters, for example, the efficiency of resource use, production processes, per unit impacts, and recycling. Such improvements are occurring for many reasons. Education is altering societal values; more consumers are recycling goods (such as newspapers and bottles) and conserving energy (such as household electricity); norms among some consumers are evolving (such as forgoing a fur coat for moral or environmental reasons); and eco-labeling programs (such as for timber and seafood) and eco-markets (such as for organic beef) are expanding. New technologies (such as catalytic converters for cars or cooling systems for refrigerators) and more efficient production (such as just-in-time assembly lines) are also reducing impacts. Corporate jockeying for market shares and profits and, to a lesser extent, policies like corporate social responsibility are also advancing environmental efficiencies (particularly for higher-end manufacturing sectors like automobiles and refrigerators). Many other forces are shifting ecological shadows, too. Government regulations—not just for the environment, but also for trade and investment—play a key role. So do pressures from nongovernmental organizations like Greenpeace or the WWF (World Wildlife Fund / World Wide Fund for Nature), international agreements to prevent ozone depletion or protect endangered species, and international aid from organizations like the World Bank and Global Environment Facility.

The evidence from this book's five cases is unequivocal: ecological shadows *do* shift, wane, even fade away. But all five also reveal that the *incremental* advances under today's current forms of environmental management are failing to prevent irreparable damage to the global environment. The cumulative progress is not keeping pace with the impact of rising consumption in a globalizing economy of ever more economic growth and ever more people. Across all of the cases, short-term economic and political factors tend to slow the speed of change. This does not mean a particular pattern of consumption in a particular location never changes appreciably for political, scientific, legal, economic, environmental, or health reasons. There are in fact countless examples arising from consumer boycotts, scientific discoveries, government bans, market crashes, and corporate bankruptcies. On occasion, a sudden change in one location even sets off a chain reaction producing better environmental management worldwide.

Still, stepping back reveals a *global* process of change that is failing to stop the environmental crisis from escalating. It also reveals that many "environmental advances" are permitting—and sometimes even causing—shadows of consumption to intensify. This process of change can also reinforce the tendency of global trade, multinational corporations, and global financing to deflect environmental costs to places and people with less economic and political power and less capacity to adapt, on the one hand, and to transfer benefits to consumers with more political and economic power, on the other.

Such a process of change, as the five cases will document in detail, has many damaging consequences. It tends to aggravate inequalities across and within states, with some people awash in excess and others bereft of the necessities to survive. It tends to allow multinational companies to expand sales in poorer countries during a phasedown of "suspect" products in wealthier ones. It tends to make it difficult to track and assign responsibility, leaving states and firms and consumers less accountable for environmental damage. It tends to expose future generations to health risks as firms pursue profits and states pursue economic growth. And it tends to cause unpredictable spillover effects across time and ecosystems.

As with lead poisoning, the thinning of the ozone layer, climate change, and rising cancer rates, the consequences of consumption can take decades, even generations, to develop. The five cases show that policy makers need to address consequences having no clean lines of causality and to take precautionary steps against effects flowing through complex systems with unpredictable outcomes. The cases also show that states should *not* assume that free trade and capitalism will significantly diminish ecological shadows, much less do away with them. Rather, international rules and institutions need to guide globalization to prevent double standards for multinational corporations, to tighten controls over the ecological impacts of trade, and to ensure that global financing supports sustainability. They show that consumers need to act locally and governments globally—at the same time. And, finally, they show that environmentalism needs to be transformed to promote more balanced personal consumption and a more balanced global political economy.

Let's turn now to the first case—the automobile—arguably, the most harmful consumer product ever for people's health and safety and for the stability of the earth's environment, yet one under relatively few international controls.

I

Automobiles

3

Accidental Dependency? The Road to an Auto World

At the beginning of the nineteenth century, cars were playthings of the wealthy. Now they speed through every culture, with the total number on the roads climbing to a billion in another decade. The global dependency on automobiles for transportation is no accident. It can be traced to the genius of entrepreneurs like Henry Ford and Alfred Sloan at General Motors, who vastly expanded markets by reducing profit margins, lobbying policy makers, advertising new models, designing cars for "obsolescence," and destroying alternative forms of transport, such as the electric trolley. The history of the automobile shows how, over several generations, as technologies develop and personal incomes rise, societies can become dependent on a consumer product. It also shows how, subject to little governmental control, this dependency can reorient communities and economies, leaving few to question the costs and risks of the resulting ecological shadows.

The first community's reaction to the first traffic "accident" is both the beginning of our indifference to the shadow effects of automobiles and a preview of the international community's reaction to today's global traffic crisis.

The Accident

Bridget Driscoll, a rather ordinary woman of her times, could never have imagined her name would appear in so many encyclopedias more than a century after her death. She became a footnote in the history of road traffic when, on a muggy August afternoon in 1896, Arthur Edsall ran her down in front of London's Crystal Palace in a demonstration "motorcar." Dr. Charles Edwin Raddock rushed out of the Crystal Palace to assist her. But her brain was "protruding," and, within moments, she became the very first person ever to die in an automobile "accident."

The inquest into her death, teeming with barristers and solicitors, probed many questions. Was the car speeding? Was Mr. Edsall reckless? Did he kill Mrs. Driscoll? Was he at fault? Some witnesses felt that Edsall was speeding when he swerved to the right to pass two cars just ahead. Other witnesses, though, felt Mrs. Driscoll was at least partly to blame, becoming confused—"rattled," in the words of one witness—by her own efforts at "dodging."

May Driscoll, at her mother's side at the time of the crash, disputed this, testifying that Mr. Edsall didn't even seem to know how to drive. Mr. Edsall denied everything, declaring he was neither reckless nor speeding, and swearing he rang the car bell and shouted out to warn Mrs. Driscoll. He wasn't even sure, he added, how she was knocked over, claiming the car stopped 2–3 inches from her body. The doctor who examined Mrs. Driscoll's body, however, concluded that the car must have hit her "very severely."

The coroner, Percy Morrison, listened to this contradictory testimony but, in the end, told the jury they could "come to no other conclusion than the car was properly driven, straight and slowly." After deliberating for six hours, the jury returned a verdict of "accidental death." In a legendary, although perhaps apocryphal, statement, the coroner said that he hoped "such a thing would never happen again."[1]

Today's "Accidental" Crisis

Since then, "such a thing" has killed at least 30 million and perhaps as many as 60–90 million people. Currently, traffic collisions, which most of us continue to call "traffic accidents," on average kill more than 800 people and injure 34,000 every six hours—the time it took the Driscoll jury to return its verdict. And traffic deaths and injuries are rising steadily. Annual deaths are likely to nearly double by 2020 as the number of cars and commercial vehicles continues to climb.[2]

This growth in the number of vehicles is putting a severe strain on the stability of the global environment. The auto industry consumes a major portion of the world's natural and energy resources. Automobiles also emit carbon dioxide, sulfur dioxide, nitrous oxide, and particulates into the air, contributing to climate change, acid rain, and unhealthy levels of smog. Even in the United States, with some of the world's strictest tailpipe emission standards, transportation accounts for a third of the nation's carbon dioxide emissions, 15 percent of the nitrous oxide emissions, and 40 percent of the volatile organic compounds.[3]

Why have automobiles proliferated so? What forces have been chang-ing the environmental and safety standards of passenger cars and trucks over the last 100 years? What is the impact of automobiles on living conditions in developing and developed countries? The answers must begin with the history of how automobiles came to be mass-produced in the United States.

Rise of America's Auto Culture

Many brilliant minds invented the motorcar—from the fifteenth-century theoretical design by Leonardo da Vinci of a self-powered vehicle to the first gasoline-powered automobile built by engineer Karl Benz in 1886. Until the year Bridget Driscoll died, motorcars were built methodically, each one different, at great expense. Then, in 1896, the American Duryea Motor Wagon Company built 13 nearly identical Motor Wagons. This was the beginning of mass production of automobiles.

After opening his Highland Park plant in 1910 Michigan, Henry Ford took mass production much further, developing an increasingly efficient moving assembly line, which would churn out millions of Model T Fords in the decades to come. It took Ford 13,000 workers to produce 260,720 cars in 1914, while it took the rest of the industry 66,350 workers to produce 286,770 cars—or about five times as many workers to produce basically the same number of cars. Prices began to plummet after the first Model T was sold in 1908—from about $850 in 1910 to $360 in 1916 to $260 in 1921—and, before long, as Ford lowered the price of his Model T to half that of most competitors, the average American worker could afford a car.[4]

Ford strove at every turn to cut costs and increase net profits, working from the principle of low profit margins on ever higher sales volumes. As he lowered his prices to increase his market share, profit per car fell from $220 in 1909 to $99 in 1914, the year he introduced a minimum wage of $5 for an eight-hour day for his factory workers. He did this, not because he was a socialist, as some were hollering at the time, but to decrease the high turnover of workers at his factory and thus save on hiring and training. "The payment of five dollars a day for an eight-hour day," Henry Ford explained afterward, "was one of the finest cost-cutting moves we ever made."[5] The results were nothing less than spectacular. Sales soared and with them overall profits, doubling from $30 million to $60 million from 1914 to 1916.

In 1916, the year sales surpassed 500,000, an average Ford assembly line worker could buy a Model T with less than 3 months' wages. By 1920, over half of the world's cars were Model Ts, and, by 1925, an average American worker—not just a Ford worker—could buy one with less than 3 months' wages. By 1924, the Highland Park plant had rolled out 10 million Model Ts; by 1927, when production ended, the number had reached 15 million.

The American car culture began to form during the first quarter of the twentieth century. The first speeding ticket was issued in 1902. The first car was stolen in 1905. Painted center lines appeared by 1911; stop signs and traffic lights by 1914, the year after more cars were made than horse buggies. By the time car radios first appeared in 1929, there was one registered automobile for every five Americans—up from one for every thirteen in 1920.[6] With so many cars came the demand for more streets and highways. The U.S. Federal Highway Act of 1921 subsidized state highway departments; meanwhile, states began to impose gasoline taxes to finance new roads. The network of roads spread rapidly. Rising car sales became a sign of economic prosperity and, by association, effective political management. In the 1920s alone, the number of passenger cars registered in the United States nearly tripled, from 8 million to 23 million.[7]

Automobile sales in the United States slowed during the Great Depression of the 1930s and World II in the 1940s. Nevertheless, the United States remained at the core of the global auto industry. Around 25 million motor vehicles (cars, trucks, and buses) were registered in 1934, some 70 percent of the global total. By 1937, the number had climbed to nearly 30 million, still around 70 percent of the global total. The number of passenger cars, trucks, and buses in the United States grew only slightly over the next decade, from about 32 million in 1940 to 33 million in 1946.[8] Sales took off again, however, in the 1950s, partly because middle-class families left the cities for the new suburbs, which had poor or no public transportation. "The automobile," in the words of economist Richard Porter, "made suburbia possible, and the suburbs made the automobile essential."[9]

By 1950, the number of registered vehicles had risen to 49 million. Drivers began to travel longer distances, especially after the 1956 National System of Interstate and Defense Highways Act funded a 40,000-mile web of free interstate highways. Everything to do with cars—whether production, advertising, renting, selling, or driving—continued to grow from the 1950s onward, becoming an ever greater part of the U.S.

economy. The rise in the sheer number of vehicles is indicative. In 1960, nearly 74 million motor vehicles were registered; by 1965, the number had surpassed 90 million.[10] In 1970, some 108 million motor vehicles were registered in the United States (89 million passenger cars and 19 million trucks and buses), or roughly one vehicle for every two Americans. Although it now accounted for only 44 percent of total registrations worldwide—a sizable decrease from the 70 percent share in the mid-1930s—the United States was still by far the largest national market.[11]

The number of cars, trucks, and buses registered in the United States continued to climb through the 1970s and 1980s—from about 133 million in 1975 to 156 million in 1980 to 172 million in 1985 to 188 million by 1991—when one out of every seven workers was somehow linked to automobiles. The largest television advertising expenditure in 1990 was on automobiles (17.6 percent of the total). Just the production of automobiles alone accounted for 3.3 percent of America's gross national product (GNP) in 1991.[12]

The number of registered vehicles rose through the 1990s, surpassing 200 million in 1995 and 213 million in 2000.[13] By 1998, the U.S. automotive sector was spending more than $14 billion on advertising alone, ranking first in advertising outlays, ahead of the general retail sector at $11.6 billion and the movie and media sector at $4.1 billion.[14] The United States had become, with only a touch of hyperbole, a mad car culture by the end of the twentieth century.

Half of Americans now live in suburbs, where motor vehicles are the primary mode of transportation, as, indeed, they are even in the cities. Public transit accounts for only 2 percent of all urban trips (compared to 7 percent in Canada and 10 percent in western Europe). Even today, with over 230 million registered passenger cars and commercial vehicles for less than one-twentieth of the global population, the United States accounts for over one-quarter of the world's total number of automobiles. Most families own two or more vehicles; private motor vehicles outnumber driver's licenses. Americans consume, as a result, over 40 percent of the world's annual production of gasoline. And the current trend is toward larger, heavier, less fuel-efficient vehicles, with sport-utility vehicles (SUVs) and other light trucks now accounting for nearly half of all vehicle purchases—a reflection, in part, of generous federal business tax deductions in recent years. SUVs and light trucks now constitute over 40 percent of the total U.S. vehicle population.[15]

Moreover, each car, SUV, and light truck in the United States tends to travel farther than in many other countries—10 percent farther, on average, than in the United Kingdom, around 50 percent farther than in Germany, and nearly 200 percent farther than in Japan. The total distance cars travel in the United States exceeds the combined total distance traveled by cars in all other developed countries. Thus cars and light trucks, which account for about 40 percent of American oil use, are a core source of global carbon emissions, contributing about the same amount as the entire Japanese economy.

Americans spend hundreds of billions of dollars on buying and operating these cars—on fuel, tires, tolls, registration fees, insurance, and repairs. These direct costs, moreover, are just a portion of total societal costs. Taxpayers subsidize motor vehicles with free parking spaces, roadwork, bridges, traffic enforcement, environmental cleanup, and the procurement of oil. To this must be added the financial costs of traffic collisions. At a rate of nearly one collision every three seconds, and 300 deaths or injuries every hour, the total costs of traffic collisions—from medical treatment to insurance to disability to police and legal services—exceeded $230 billion ($820 per person) in 2002.[16]

Cars have become "essential" in the United States for many reasons. Thanks in part to advertisers, owning a car has come to mean independence, success, and status; it also evokes coming of age, a desire for speed, a sense of adventure and romance. But cars have also become "essential" because they are in fact a necessary, or at least a highly convenient, part of everyday life in the United States, whether to commute between home and work, run errands, transport children, or take the family on weekend outings.

Accidental Dependency?

Such dependency on the automobile is not accidental. The true genius of entrepreneurs like Henry Ford—who died in 1947 worth $1 billion, ranking as one of the wealthiest business people of all time—was making money by expanding markets with low profit margins and high sales volumes. New markets arose in part because more workers wanted—and could now afford—a car and because government policies put in place the infrastructure to support more cars. Other creative entrepreneurs, like Alfred Sloan, president and later chairman of the board of General Motors from 1923 to 1956, expanded markets with strategies like "planned obsolescence," which encourages customers to upgrade to a new model every few years. Sloan stimulated consumption as well by introducing "style" into automobiles, selling models with differing features and offering credit to buyers. Such entrepreneurs also expanded markets by destroying, or at least contributing to the decay of, other modes of transportation—in particular, the trolley (electric streetcar).

The trolley was cheap and convenient in many U.S. cities at the turn of the twentieth century. Nearly every city of 25,000 or more had trolleys by 1900. Passengers took some two billion trolley trips in 1890. Just 30 years later, the number of trips had jumped to 15.5 billion a year. The trolley by 1920 was, according to historian Mark Foster, "the chief mode of transit for nearly all classes and ages: businessmen commuting to their offices; industrial workers going to factory jobs; housewives on family errands; and children traveling to and from school."[17] Trolleys from the 1920s onward went into decline for a host of reasons. Some trolley systems were mismanaged by shortsighted, incompetent, or corrupt executives. Some went bankrupt, unable to adjust to the changing needs of local economies, and unable to combat the convenience and increasing affordability of cars.

Yet automakers also targeted the transit systems. One infamous example became evident in the 1949 legal case of the U.S. government versus National City Lines. A holding company formed by General Motors (GM), Standard Oil of California, and Firestone Tire and Rubber, National City Lines set about acquiring transit firms in over 40 cities. It then demolished the trolley lines and replaced the trolleys with GM buses, which ran on Firestone tires. The federal court in Chicago found General Motors, Standard Oil of California, and Firestone Tire and Rubber guilty of antitrust violations. Other firms in the automotive sector undercut transit firms by granting exclusive (and sometimes illegal) supplier contracts, by providing financial and technical assistance to municipalities to switch from rail to road transit systems, and by lobbying and financing supportive politicians. The result was the loss of rail transit systems—like the 1,100-mile Pacific Electric Red Car System in Los Angeles and Orange Counties—and the increasing reliance of individuals and economies on cars and buses. Over 90 percent of America's trolley system was dismantled by the 1950s.[18]

It's too simple to depict the victory of autos and buses over rail transit systems as a corporate-government conspiracy. Many cases exist where local governments in the United States have supported rail transit systems. And in a few cases, too, public protests have blocked the growth of road systems since the late 1950s (for example, when citizen groups halted freeway construction in San Francisco, Boston, and Washington, D.C.). Yet, as sociologists Peter Freund and George Martin write, the reality is not far off a conspiracy for a simple reason: "The auto-industrial complex is monied and quite influential; the interest groups that oppose it are considerably less monied and less influential."[19]

As the previous sections have shown, the American auto market has continued to expand since the days of National City Lines. So have the auto markets of western Europe and Japan, where governments and firms similarly promote the growth of automobile consumption. The three markets, although slowly declining as a percentage of the global share, still account for around 50 percent of global production of motor vehicles and about 65 percent of registered motor vehicles.[20] As chapter 4 will document, however, these markets have also been changing in other ways over the last half century, ways that include adopting much higher environmental and safety standards.

4

A Better Ride: Selling Safe and Clean

Government regulations and competition among automakers since the 1960s have made the typical automobile safer and cleaner, in wealthier and poorer states alike. Governments across the First World began passing legislation in the 1960s and 1970s to establish standards for air pollution and auto emissions. California has been a leader in this regard since 1966, when it set the world's first standards for carbon monoxide and hydrocarbons from automobile tailpipes. It has progressively strengthened regulations for automobile emissions, a trend seen in other jurisdictions in North America, Europe, Japan, Australia, and, to a lesser extent, in developing countries as well. Over this time, governments have also passed laws and set up programs to make roads and cars safer by, for example, requiring seat belts, enforcing speed limits, and mandating infant or child car seats. Again, as with regulations and laws to protect the environment, those for traffic and vehicular safety are more far-reaching and better enforced in the First World than in the Third World.

To meet stricter regulatory standards and jockey for markets, automakers have been upgrading new models with environmental technologies like catalytic converters and fuel injection systems as well as with safety features like antilock brakes, safety glass, and air bags. Because older or badly maintained automobiles account for much of vehicle emissions overall, governments have in turn encouraged consumers to maintain their older cars and trucks or trade them in for newer ones. Some have worked with automakers to improve the recycling of junked vehicles, having reasonable success in the First World with materials that are easier to recycle, such as iron and steel, and less success everywhere with materials that are harder to recycle, such as plastic and "fluff" (hazardous residue from shredding a vehicle). Even here, however, governments in Europe and Japan are now beginning to improve recycling rates by making automakers responsible for recycling, which motivates them to design vehicles with cost-effective recycling in mind, such as labeling plastic components with content codes.

As a result of these regulations and technologies, and as this chapter documents, the shadow effects of individual automobiles on the lives of people and on the quality of ecosystems have been declining steadily for half a century. A typical new passenger car in the United States, for example, is far less polluting than a 1960s model. Traffic fatalities per driver *and* per mile driven have been falling across the developed world since the 1960s as well.

Such progress shows how domestic legislation and corporate advances can combine to improve the environmental performance of consumer goods and how trade and investment can then transfer these improved products into jurisdictions with lower environmental regulations. That said, such progress has not, as chapter 5 will argue, kept the ecological shadow of automobiles *as a whole* from intensifying and shifting into ecosystems and onto people less able to resist or adapt.

Regulating Traffic in the First World

Governmental and corporate efforts in the First World to improve the environmental performance of automobiles gathered momentum from the 1950s to the 1970s. The U.S. federal government, for example, gradually strengthened air quality legislation, passing first the Air Pollution Control Act of 1955 and then the Clean Air Acts of 1963 and 1970, which set new standards and provided more funding for research on air pollution.[1] Many other developed countries—Japan, Britain, Germany, Canada, and Australia, to name just a few—also took regulatory steps to improve the environmental performance of vehicles in the 1970s, aided by the quadrupling of oil prices, which created a sudden global demand for more fuel-efficient cars. Because the "use phase" accounts for nearly 90 percent of a motor vehicle's consumption of energy during its life cycle, advances in its environmental performance are crucial for energy conservation.[2]

California has played a leading role in raising the bar for environmental standards for automobiles not only in the United States but also in Europe and Japan. By the early 1960s, California was requiring the use of positive crankcase ventilation, a technology to control hydrocarbon emissions. In 1966, it became the world's first jurisdiction to set standards for automobile tailpipe emissions for carbon monoxide and hydrocarbons; that year, the California Highway Patrol began random roadside inspections of smog control devices. Three years later, California became the first U.S. state to pass ambient air quality standards for total suspended particulates, nitrogen dioxide, photochemical oxidants, sulfur dioxide, and carbon monoxide. And, in 1971, it became the first state to set automobile nitrogen oxide standards.

Over the next four decades, California continued to strengthen automobile regulations. The California Smog Check Program went into effect in 1984, mandating inspection of automobile emission control systems every two years. In 1988, California passed regulations requiring onboard computers to monitor emission performance by the model year 1994. In the 1990s, it enacted legislation to require cleaner diesel fuel and phase in cleaner-burning gasoline, and, in 2001, it set new standards to reduce diesel soot and smog-forming emissions by 90 percent from large diesel engines, such as those in big rig trucks, by the model year 2007.[3]

Although the regulatory changes have created incentives for automakers to research, design, and market cars with better environmental technologies, as some firms have seen potential competitive advantages in them, the advances themselves have lowered resistance to the new rules. Because the advances are too extensive to review fully, the next section touches on just two of the most notable ones—the catalytic converter and fuel injection.

Converters and Injection

Catalytic converters, which scrub automobile exhaust of pollutants causing smog, first appeared on new vehicles in the early 1970s. These spread quickly through the North American and Japanese markets. (Most of Europe didn't widely adopt catalytic converters until the early 1990s.) After the mid-1970s, nearly all new cars sold in the United States had catalytic converters. Automakers began to compete to install more effective converters. By 1977, Volvo was marketing a car as "smog-free"—with the first three-way catalytic converter to control hydrocarbons, nitrogen oxides, and carbon monoxide. By the early 1980s, more efficient catalytic converters were enabling some U.S. firms to exceed some federal emission standards to reduce smog.[4]

Fuel injection improves fuel efficiency, lowers emissions, and increases the power of gasoline engines. Since fuel injection systems first appeared in the 1950s, auto firms have been steadily perfecting them, introducing electronic fuel injection in 1967, which significantly improved environmental performance. Fuel injection was common on automobiles in Europe by the 1980s and standard on new models in the United States by 1990. (The 1990 Subaru Justy was the last model sold in the United States with a carburetor.)

Rising Emission Standards: California as Exemplar

Stronger regulations and better technologies like the catalytic converter and fuel injection have steadily decreased the environmental impact of individual automobiles in most developed countries over the last four decades. A typical new automobile in the United States, for example, is over 95 percent less polluting for emissions such as nitrogen oxides, carbon dioxide, and hydrocarbons than one sold in the 1960s.[5] Such progress has, in some cases, allowed the net ecological impact of automobiles to decline even when the number of vehicles on the road is rising rapidly. The data from California are especially impressive. In 1970, there were just over 12 million registered motor vehicles in California; annual vehicle miles traveled (VMT) totaled 110 billion. The average emission of nitrogen oxides per vehicle was 5.3 grams per mile; for hydrocarbons, it was 8.6 grams per mile. The vehicle emissions for nitrogen oxides and hydrocarbons were nearly 1.5 million metric tons per year.

A decade later, with stricter emission regulations and better environmental technologies for vehicles, the average emission of nitrogen oxides per vehicle was down to 4.8 grams per mile, while, for hydrocarbons, it was down to 5.5 grams per mile. Total vehicle emissions for nitrogen oxides and hydrocarbons remained at the 1970 level even though there were now five million more vehicles with 45 billion more vehicle miles traveled. By 1990, 23 million vehicles were registered in California; vehicle miles traveled in that year totaled 242 billion. The average emission of nitrogen oxides per vehicle had fallen to 3.0 grams per mile; for hydrocarbons, it was 2.7 grams per mile. Total vehicle emissions for nitrogen oxides and hydrocarbons were down about 200,000 metric tons per year from the 1980 level despite an increase of 87 billion vehicle miles traveled; they would continue to decline over the next decade even as the total number of cars continued to climb.

By 1995, the average emission of nitrogen oxides per vehicle was down to 2.2 grams per mile, while the average emission of hydrocarbons was down to 1.8 grams per mile. The total annual amount of nitrogen oxides and hydrocarbons from auto emissions in 1995 was approximately 1 million metric tons, one-third less than the 1970 level even though total vehicle miles traveled were 146 percent higher. (The 26 million registered vehicles traveled a total of 271 billion miles in 1995.)

Five years later, the California averages for emissions of nitrogen oxides and hydrocarbons were again down, although just by a fraction this time—another 0.1 grams per mile for nitrogen oxides and 0.2 grams per mile for hydrocarbons. Total vehicle emissions for nitrogen oxides and hydrocarbons rose somewhat—to around 1.1 million metric tons per year—with vehicle miles traveled rising to 280 billion per year. This higher cumulative figure for 2000, however, is still around 200,000 metric tons per year less than in 1990 and 400,000 metric tons per year less than in 1970.[6]

Other cities in the First World, even ones with reasonable air quality, are following leaders like California to raise vehicle emissions standards. In Vancouver, for example, a program called "Air Care" to test whether light-duty vehicles meet minimum standards for tailpipe emissions of hydrocarbons, carbon monoxide, and nitrogen oxides decreased harmful vehicle emissions by 35 percent from 1992 to 2002. Cleaner fuels and new vehicle technologies lowered emissions by another 31 percent over this same period. It's reasonable to expect air quality in Vancouver to continue to improve, in part because Canada as a whole began phasing in even stricter emission standards for new vehicles in 2004, aligning Canadian emission standards with American ones. These will eventually reduce allowable emission levels by up to 95 percent. Canada has been taking steps to ensure cleaner gasoline, too. Regulations now prohibit the sale of gasoline with more than 1 percent benzene (by volume) and require that it have an average sulfur concentration of no more than 30 milligrams per kilogram (about 38 milligrams per U.S. gallon).[7]

Recycling the Junk

Governments in First World countries such as Canada are also providing incentives for owners to trade in old vehicles, a logical way to further lower emissions in Canada where older or poorly maintained vehicles (which comprise 10–15 percent of the vehicle fleet) account for up to 50 percent of total vehicle emissions. On the other hand, trading in older vehicles adds to a growing challenge for all states: how to safely dispose of car parts and effectively recycle them.

Some governments, like Japan's, are imposing new regulations to increase recycling rates for automobiles. Japan's current recovery and recycling rate for vehicles is between 75 and 85 percent of total weight.

The government aims to raise this to 95 percent by 2015. The European Commission, under its end-of-life vehicles (ELVs) directive, is now requiring manufacturers and importers to set up scrap yards to take back, "de-pollute," and recycle used vehicles bearing its logo. Around 14 million vehicles reach the "end of life" in the European Union (EU) each year. The ELVs directive aims to establish rules and create incentives so that manufacturers design cars with recycling in mind. Provisions thus require automakers to code parts and materials and to provide dismantling information. Like Japan, EU member states are seeking to reuse, recover, and recycle 95 percent of vehicle weight by 2015.[8] They're seeking as well to solve the problem of owners simply abandoning old vehicles rather than paying for their disposal.

The auto-recycling industry in North America, like the one in Japan, recovers between 75 and 85 percent of the total vehicle weight. And, as in Europe and Japan, most of this is iron and steel. The steel-recycling rate for cars in the United States—calculated by comparing the annual amount recycled with the annual amount used to manufacture new vehicles—was well over 90 percent even in the 1990s. Today, recyclers are retrieving slightly more steel from scrapped vehicles than manufacturers are using for new vehicles (in part because some newer cars contain less steel). Not all of this recycled steel goes back into car manufacturing. Indeed, only about one-quarter of the steel in new car bodies is recycled steel, although the proportion is higher for internal steel parts.[9]

Automobiles are one of the world's most recycled commodities, with 95 percent of vehicles in North America and Europe ending up with dismantlers and shredders. Dismantlers will salvage parts in decent condition, including engines, tires, batteries, fuel, catalytic converters, and air bags, but much of their efforts are focused on recovering steel, a relatively easy process. In contrast, recycling fluids in radiators, brake systems, and transmissions or toxic metals like mercury, lead, and cadmium is both expensive and laborious, as is recycling nonmetallic car parts made from rubber, glass, or plastic. Metallic car parts account for between 70 and 75 percent of a typical vehicle's weight. Of the nonmetallic parts, tires and elastomers constitute about one-quarter, plastics around one-third, and glass about 13 percent.[10]

Recycling Plastic Cars

The proportion of plastic in cars, in everything from bumpers to door lining to contoured upholstery, has been steadily rising for decades. It's light, cheap, and flexible, increasing fuel efficiency and design possibilities. Plastic climbed from 0.6 percent of vehicle weight in 1960 to 7.5 percent in 2000. This may not seem like much, but it's equal to 4.3 billion pounds of plastic per year in the United States alone.[11] The recycling of plastic car parts, however, though gradually increasing in developed countries, lags far behind the recycling of steel.

Dismantlers in the United States, for example, still manage to retrieve only a small fraction of the plastic in automobiles. This generally becomes shredder residue or "fluff"—a mixture of plastics, rubber, glass, paints, dirt, oils, and miscellaneous pieces left over after a giant magnet extracts metals from the crushed vehicle hulk. Shredder fluff is considered hazardous waste by the international community and many national governments because of the polychlorinated biphenyls (PCBs) and other hazardous substances it can contain.

The United States produces about 5 million metric tons of this fluff every year. Most of it ends up in landfills or incinerators. One reason for the low rates of recycling car plastics is the expense and difficulty of sorting and processing the more than 20 types of plastic in an average vehicle, and the even greater difficulty of doing so for older vehicles whose plastic parts are not stamped with their composition and grade. Another reason is the technical difficulty of recycling small amounts of low-grade plastic waste. And a third, perhaps most important, reason is the lack of strong markets for recycled car plastics and shredder residue.[12]

A few scientists are investigating imaginative solutions to reduce plastic waste. One novel idea is to make plastic car parts that will biodegrade upon disposal. In this regard, the Ecology Center, an American NGO advocating for clean air, safe water, and healthy communities, rates Toyota as the "clear sustainable plastics leader"—well ahead of Honda, DaimlerChrysler, Ford, Nissan, and General Motors. Toyota is developing biodegradable plastics from renewable materials such as corn and sugarcane and working to eliminate polyvinyl chloride (PVC), a common plastic that releases toxic chemicals such as dioxins, furans, and PCBs, from its vehicles.[13]

Regulating Traffic Safety in the First World

Not only vehicles but also drivers and passengers in developed countries have become much safer over the last 30 years. There are a host of reasons for this. Activists such as consumer advocate Ralph Nader, who wrote the 1965 bestseller *Unsafe at Any Speed*, raised public awareness of the need for higher safety standards within developed countries. Many governments passed new laws to establish baseline standards, such as the U.S. National Traffic and Motor Vehicle Safety Act of 1966. Many began to design and maintain safer roads; some put in place speed bumps and more appropriate speed limits and began to enforce penalties for speeding and drunk driving. Some governments also began to require that automakers equip their vehicles with safety features. One of the first and most successful was the seat belt.

Buckling Up and Braking

Although a few physicians put lap belts in their own cars and called for them in all new cars as far back as the 1930s, it wasn't until the 1950s that auto manufacturers, including Ford, Chrysler, and Volvo, first included belts in some models. These became increasingly common—and more elaborate—in the 1960s. Some jurisdictions, like the Australian states of Victoria and South Australia in 1964, began to require front seat belt anchorages on new cars. Victoria was the first to pass a law on mandatory use of restraints in 1971; that year, the number of traffic fatalities among passengers and drivers in the state fell by 18 percent.

Soon others followed Victoria's lead. By 1972, all Australian states required seat belt use, as did New Zealand and West Germany. Yet, for a long time, many did not. The United Kingdom, for instance, didn't require front seat belt use until 1983. Under the new law, front seat belt use went up 58 percent—from 37 to 95 percent—while hospital admissions for traffic injuries went down 35 percent.[14] Today, most governments in the First World require seat belts. Numerous studies show this saves many lives, reducing traffic fatalities among drivers and front seat passengers by 40–65 percent.

Many other advances have made cars and trucks safer, too. The drum brakes of 50 years ago could require 100 pounds of pedal force to work when hot. Power brakes, common by the late 1950s, require far less pedal force no matter their temperature. Split-brake systems, which retain partial braking power even in the event of a leaking brake line, first appeared in the early 1960s and were mandated by some nations, such as the United States, by the late 1960s. Today, technologies like the antilock braking system (ABS) are even more effective and reliable.

Other safety features, such as air bags, child or infant car seats, and safety glass, are also reducing fatalities from traffic collisions. Windshields that would break without shattering appeared in 1953, although they could still be deadly if a person's head crashed through them. The first "high-penetration-resistant" windshields appeared in 1965; just three years later, the United States required these on all cars. Air bags first appeared in the mid-1970s. By 1999, the United States required dual air bags on all new passenger vehicles. These safety features significantly reduce the chances of dying in a traffic collision. Studies show, for example, that air bags can reduce driver deaths by 22–29 percent in front-end crashes, while safety car seats can lower infant deaths in car crashes by 71 percent and small child deaths by 54 percent.[15]

In the latest wave of safety features, automakers are developing "intelligent" vehicles with new safety technologies that range from the simple (audible seat belt and speed reminders) to the sophisticated (automatic notification systems to alert emergency crews after a crash). The near future could well see cars with radar-equipped computers able to override drivers and avoid collisions by directing their vehicles to swerve and brake.

The SUV Setback

Four out of ten Americans polled by the Associated Press in 2003 thought that it was safer to ride in an SUV than in a car. The tendency of SUVs to roll over, however, can make these vehicles as dangerous as cars for the driver and passengers—or more so. Thus, in the United States, the National Highway Traffic Safety Administration calculated that, for 2001, occupants in SUVs were 3 percent more likely to die in a collision than occupants in cars (162 deaths per million SUVs versus 157 deaths per million for cars); three years later, the Department of Transportation calculated that figure at nearly 11 percent.

The evidence that SUVs increase the risk of death for other drivers of cars is even stronger. The National Highway Traffic Safety Administration estimates that when an SUV, rather than an ordinary car, hits the driver's side of a car, the other driver is almost 5 times more likely to die.[16]

That said, a few basic statistics on traffic fatalities clearly show that traffic conditions in developed countries are safer overall, even after discounting for higher survival rates from better medical and emergency care.

A Safer Drive in the First World

Traffic fatalities in the United States fell from more than 25 per 100,000 in 1966 to 15 per 100,000 in 2000; those in the United Kingdom fell from 15 per 100,000 in 1966 to around 6 per 100,000 in 2002; and those in Australia, from 30 per 100,000 in 1970 to fewer than 10 per 100,000 in 2001. Drivers and passengers are also much safer over longer distances. A person's chances of dying in a vehicle in the United States, for example, were over four times higher per mile driven in 1953 than they are today.[17]

On balance, then, the conclusion is clear: the modern automobile is much safer and contains better environmental technologies. Such changes certainly benefit consumers and advance environmental management. Yet, as chapter 5 will show, rising numbers of cars and light trucks on the roads of the developing world are overwhelming the benefits of better-made vehicles.

5

The Road Tolls

A 2008 Toyota Prius no doubt causes less ecological harm than a 1965 Toyota Corona; and a 2008 Ford Taurus is no doubt far safer than a 1965 Ford Mustang. Still, the shadow effects of today's safer and cleaner automobiles are growing as the numbers of these vehicles rise across the globe. A company like Toyota is now a model of how to produce large outputs with just-in-time efficiency, less waste, and high quality controls. It's at the forefront of research in environmental technologies, too. But Toyota's ultimate goal is not just to expand sales, capture markets, and increase profits—or to be a pioneer in eco-friendly automotive engineering—but also to garner the prestige of surpassing General Motors as the world's number one automaker.

This global competition to produce more has put well over 800 million cars and commercial vehicles on today's roads—up almost twelvefold from 70 million in 1950. With automakers marketing sport-utility vehicles and other light trucks as passenger cars, many of the newer automobiles are big and heavy (thus more dangerous for ecosystems and pedestrians). Quickly rising sales in the Third World means this region accounts for an increasing share of the global vehicle fleet, as firms like GM, Ford, and Toyota rush into booming markets like China.

Driving these cleaner automobiles uses up 15 percent of the world's energy output and generates one-fifth of the world's energy-related carbon dioxide emissions every year. Junking used vehicles for new models also adds to the global ecological burden, especially in developing countries without effective recycling facilities. Meanwhile, traffic collisions of these safer automobiles now kill more than 3,000 people and injure 137,000 *every day*. Even with all of the advances in safety over the last half century, the chances of dying in a traffic "accident" remain significant in the First World. Yet the shadow effects of the expanding auto industry are far greater in developing countries, where traffic laws and infrastructure are inadequate. Already, 85 percent of traffic deaths now occur in these countries, with children facing particularly high risks. And the death toll rises every year, even as it falls in the First World.

Producing More Traffic

The number of passenger cars and commercial vehicles registered worldwide has been rising steadily over the last five decades, from 127 million in 1960, to 246 million in 1970, to 411 million in 1980, to 583 million in 1990, to 735 million in 2000, to more than 825 million today.[1] The global auto industry is now producing around 67 million light vehicles—passenger cars, station wagons, and light commercial vehicles under 6 metric tons (13,200 pounds)—each year, an increase of 10 million or so from 2000. Although millions of vehicles are junked or recycled each year, industry analysts predict the overall number of vehicles will continue to grow rapidly, with somewhere between 2 and 3.5 billion light vehicles on the world's roads by the middle of this century.[2]

Toyota's growth has been especially impressive since 2000, accounting for half of the total increase in global production. Several years ago, the chairman of the board, Hiroshi Okuda, set a corporate target of capturing 10 percent of the world market—to enter what he called the "Global Ten." After achieving this, he set a new target of 15 percent of the global market to overtake General Motors as the world's top producer. In 2006 alone, Toyota's global production surged another 10 percent to more than 9 million vehicles, putting GM just 162,000 ahead. Despite increasing sales by 3 percent, GM was only 3000 ahead in 2007, with Toyota on track to become global sales leader in 2008 or 2009, a position GM has held since 1931.

The Toyota Way

Toyota's success stems in large part from a corporate culture of always striving for greater efficiency, better quality controls, less waste, steady production increases, and ultimately a lower sticker price for a more reliable vehicle. The Toyota Production System—made famous in the book *The Machine That Changed the World*[3]—aims to continually improve the manufacturing process by bringing parts together "just-in-time," thus eliminating inefficiencies, as well as concentrating on consistent high quality, such as allowing workers to halt the production line if they spot something amiss.

Toyota's system uses fewer natural resources and generates less waste per unit produced. Auto firms in other countries adopted many of Toyota's manufacturing strategies, and "in the process American and European cars went from being unreliable, with irritating breakdowns, leaks and bits dropping off in the 1970s, to the sturdy, reliable models consumers take for granted today."[4]

Toyota remains highly efficient and has recently emerged as a leader in innovative environmental technologies such as hybrid cars like the Prius, which combines a conventional gas engine and an electric motor to maintain power on the open road. During start-up, idling, or slow traffic, it relies only on the electric motor (which itself emits no greenhouse gases). As a result, the Prius achieves over 61 miles per gallon in city driving and over 57 miles per gallon on the open road. It uses half as much gasoline as a comparable new American car, emitting half as much carbon dioxide, and some 90 percent less nitrogen oxides and hydrocarbons.[5]

Although the Prius is a notable advance, at the heart of what Toyota employees call the "Toyota Way" is expansion—the drive to manufacture and sell more, and thus to win the game of being number one in terms of vehicles sold globally. The growth in the number of plants makes this clear. Toyota had 11 manufacturing plants in 9 countries in 1980; ten years later, it had grown to 20 in 14 countries. By 2005, it had expanded to 46 manufacturing plants in 26 countries, with design and engineering facilities in the United States, France, Belgium, and Thailand. Like all of the top automakers, Toyota markets its cars and trucks with sophisticated advertising campaigns. It develops new models with clockwork regularity, rushing to fill market gaps or compete with rivals. There are, for example, more than 60 models in Japan alone, with many more designed specifically for overseas markets.[6]

Producing Big Traffic

The sales success of SUVs, minivans, and other light trucks made by Toyota, General Motors, Ford, and DaimlerChrysler in large part explains the growth in auto production since the 1990s. SUVs and other light trucks constitute a growing share of total purchases and, if the current trend continues, they will account for half of the world's passenger vehicles by 2030.[7] Not only do SUVs use more fuel per mile traveled, but also take up more space than standard passenger cars, further congesting cities like Los Angeles, where roads and parking already account for two-thirds of all urban land space. Partly because of the increase in these larger, heavier vehicles, the average weight of a standard passenger vehicle in the United States, having fallen during the 1970s and 1980s, began to rise again in the 1990s, reversing many of the earlier gains in fuel efficiency from lowering vehicle weights.[8]

Over the last century, then, the automobile became integral to the economies—some even argue to the cultures—of Japan, Europe, and the United States. This century's frontier for the global auto industry is the developing world.

Paving Developing Countries with Autos

GM, Toyota, Ford, Volkswagen, Nissan, BMW, Honda, and Hyundai-Kia are now aggressively expanding markets in developing countries. Some of the sales are First World discards. Japan, for example, exported around 1 million used motor vehicles in 2003—worth about $2.7 billion.[9] But most of the sales to the Third World are new cars.

A glance into China reveals the potential for big auto sales. At the end of the 1990s, there were fewer than 10 vehicles per 1,000 people (compared to 780 per 1,000 in the United States). The total number of cars has been rising steadily since 2000—by 4 million, for example, in 2003, with over 400,000 vehicles added to Beijing's roads alone (nearly a 25 percent increase in Beijing). The International Energy Agency now predicts the period 2005–2030 will see a sevenfold increase across China, to 270 million cars and trucks. Not surprisingly, all of the major automakers are now jockeying for position in this exploding market. GM, for example, now operates seven joint ventures and two wholly owned enterprises in China. It sold over 665,000 vehicles in 2005, an increase of 35 percent from the previous year (now accounting for just over 11 percent of market share). Since 2001, China as a whole has swept past Canada, France, Germany, South Korea, and Spain to become the third-largest motor vehicle producer (after the United States and Japan).[10]

The Environmental Tolls

The future, then, will certainly see the total number of motor vehicles continue to rise. This globalization of automobile consumption is already damaging the global environment and putting millions of people—especially the world's poorest—at risk. A quick statistical overview confirms this. Over recent decades, automobiles have accounted for almost half of all the oil and rubber, a quarter of all the glass, and 15 percent of all the steel consumed each year across the globe. Passenger cars alone currently account for about 15 percent of global energy use. To this must be added the energy required for constructing and maintaining roads for these cars. The roads in turn disrupt natural hydrologic cycles, which can cause flooding and reduce groundwater supplies.[11]

Automobiles are also emitting carbon dioxide, nitrous oxide, sulfur dioxide, and particulates into the air, contributing to smog, acid rain, and climate change. Smog—a word coined from smoke and fog in 1905—is now less severe than a few decades ago in some cities like Los Angeles. But it remains a serious and growing environmental and health problem in many cities, especially in the developing world, where transport often accounts for between 70 and 80 percent of local air pollution. Globally, outdoor air pollution kills over 24,000 children every year. Even wealthy places continue to struggle with controlling outdoor air pollution. In Canada, for example, a 2005 study by the Ontario Medical Association estimates that over 5,000 people die prematurely because of air pollution—a figure the association predicts will increase to over 10,000 people by 2026.[12]

Motor vehicles are one of the biggest causes of climate change, and the impacts have been steadily increasing as the volume of traffic grows. The national impact of automobiles on global carbon dioxide emissions naturally varies across jurisdictions. Personal consumption in a place like Canada accounts for more than a quarter of the country's total greenhouse gas emissions; of these, passenger road transportation accounts for half, with private vehicles by far the most significant emitters.

Globally, motor vehicles now account for about one-fifth of total energy-related carbon dioxide emissions (and some three-quarters of total emissions from transportation). Even in the United States, with great strides in the environmental performance of vehicles, transportation continues to account for one-third of total carbon dioxide emissions.[13] Some of the technological devices to produce "cleaner" automobiles are having unexpected ecological consequences. For example, catalytic converters, which, as noted in chapter 4, effectively reduce many smog-producing emissions, produce nitrous oxide (also known as "laughing gas") as a by-product. Nitrous oxide now accounts for over 5 percent of U.S. greenhouse gases: about 15 percent arises from cars and trucks with catalytic converters (other major sources include manure and fertilizers). Catalytic converters also emit heavy metals (rhodium, palladium, and platinum), with unknown health consequences.[14]

SUVs and other light trucks are especially hard on the environment. An average SUV in North America traveling 12,000 miles (20,000 kilometers) per year emits half again as much carbon dioxide—some 6 instead of 4 metric tons—as an average mid-sized sedan. Many countries, including the United States, do not impose the same environmental restrictions on SUVs and other light trucks as on cars, with the result that, according to the Union of Concerned Scientists, an average truck sold in the United States emits "2.4 times more smog-forming pollution and 1.4 times more global warming gases than the average" car.[15]

Nor does the ecological impact of such trucks end with greater emissions. They are scarring deserts and stirring up vast amounts of dust—a phenomenon geographer Andrew Goudie has labeled "Toyota-ization" because so many of the four-wheel-drive vehicles in places like North Africa are Toyota Land Cruisers. Toyota-ization combines with deforestation and desertification to generate more frequent and intense dust storms. Professor Goudie's analysis of satellite images in Saharan Africa, for example, shows a tenfold increase in dust storms over the last half century. This has resulted in as much as 3 billion metric tons of dust blowing into the atmosphere every year, enough, Goudie calculates, to disrupt climate patterns and destroy coral reefs.[16]

Junking the increasing number of old cars and trucks creates many environmental problems as well. The capacity to recycle old vehicles in wealthy places like Japan, western Europe, and North America has been, as the last chapter documented, steadily rising (and should continue to do so over the next decade). This is not the case, however, in most of the developing world. Vehicle recycling, for example, hardly occurs in the poor countries of Africa. National legislation to require more recycling of vehicles can also shift the burden of disposal to countries without effective recycling capabilities (for example, dumping old German cars in eastern Europe or old Japanese cars in Southeast Asia).[17]

Even comparatively wealthy developing countries like China are just beginning to develop a recycling infrastructure. The number of vehicles in use in China now exceeds 50 million, with more than 2 million vehicles being junked every year. At current rates of growth, this number will rise to over 3 million by 2010. Although China passed a law regulating the disposal and recycling of vehicles in 2001, according to the Chinese government, 90 percent of vehicles reaching "end-of-life" status—vehicles that are unsafe and highly polluting—remained illegally in use as of 2004.[18]

Dying on the Roads

The increase in global traffic is killing more than just natural environments: it's also killing increasing numbers of drivers, passengers, and pedestrians in countries with poor infrastructure and inadequate safety rules. Many newer vehicles, as documented in chapter 4, are safer for drivers and passengers (especially in terms of crash tests). But some, such as SUVs, are more dangerous for occupants in other cars and far more dangerous for pedestrians. The height of the front end of many SUVs, for example, means they are over two times more likely than a standard passenger vehicle to kill or severely injure a pedestrian in a collision.[19]

Despite safer conditions for people in most developed countries, the total number of traffic deaths and injuries each year continues to climb. At least 30 million people have died in traffic collisions over the last century. Given the uncertainties of global data, the actual number could be two to three times higher. Today, traffic collisions injure between 20 and 50 million and kill almost 1.2 million every year (about one-third of these deaths are pedestrians). The WHO and World Bank predict that annual traffic deaths will exceed 2 million by 2020.[20]

An average person, even living in a developed country, still has a roughly 1 percent lifetime chance of dying in a traffic collision.[21] Not surprisingly, the risk rates vary somewhat across developed countries. In the United States, for example, the National Safety Council estimates the lifetime odds of dying in a motor vehicle collision at 1 in 84, or 1.2 percent.[22] Yet the dangers are far greater in developing countries. Today, of the total number of traffic deaths globally, 85 percent are in low- and middle-income countries. Death rates in developing countries are rising, even as rates are falling in developed states. Traffic fatalities per 10,000 people jumped from 1975 to 1998 by more than 44 percent in Malaysia, by more than 79 percent in India, by just over 237 percent in Colombia, by 243 percent in mainland China, and by nearly 384 percent in Botswana.[23]

Children in the developing world are especially at risk. Six times more children per 100,000 people die in traffic collisions in low- than in high-income countries. Indeed, poorer countries account for 96 percent of all children killed in traffic collisions. Pedestrians, cyclists, and motorcyclists account for many of the deaths in low-income countries, with fast speeds, wild road rules, and few safety features (even fairly simple ones like helmets) combining to create great hazards. Public transportation, too, is often very dangerous, as the colloquial terms for buses in Lagos, Nigeria suggest: *danfo* (flying coffins) and *molue* (moving morgues). The resulting burden on medical systems is high, with between 30 and 86 percent of all trauma admissions in low- and middle-income countries now related to traffic injuries.[24] Meanwhile, governments and automakers continue to dismiss such tragedies with empty assurances that these "accidents" will become a thing of the past with the globalization of better cars, roads, and laws—even as the yearly toll in traffic deaths climbs relentlessly toward 2 million.

6

The Globalization of Accidents and Emissions

The automobile over the last 100 years went from being a luxury toy to a normal purchase for normal people in just about every culture on earth. Hundreds of millions of people now depend on private motor vehicles. Workers rely on them to commute, parents rely on them to transport children to school and play, and families rely on them to vacation at cottages and beaches. Granted, many of these drivers *could* use public transportation, but it's generally slower and less convenient.

The structural dependency of societies on automobiles is not an accident of history. Networks of auto, tire, and construction companies have worked hard to expand roads and obstruct (or even destroy) public transit. Governments have subsidized the use of private vehicles with, for example, free roads and parking space. Manufacturing and advertising firms have associated "owning a car" in the minds of consumers with sex, freedom, speed, adventure, and affluence. One result is cars now move through popular cultures with relatively few institutions or consumers ever questioning the costs and risks of the ecological shadows of automobiles.

Over the last century, the most influential forces changing the environmental impacts of automobiles have been technological advances, government regulations, and corporate positioning. Starting with industrialists like Henry Ford, automakers have competed to produce more cars with less labor, resources, time, and waste. As output rose with less input, cars became cheaper and sales went up.

Ford's Model T epitomizes this first era of rapid expansion. Putting the Model T on an assembly line, Ford was able to use savings from more efficient manufacturing to lower prices so average American workers could afford the luxury of independent travel. Ford sold his 1908 Model T for $850; thirteen years later, it cost just $260. His production and marketing strategies were so successful that half of the world's cars were Model Ts by 1920. By the mid-1920s, Ford had sold 10 million Model Ts; another 5 million were sold over the next few years. The 1920s was an era of fast growth in the American car industry, and, by the end of the decade, over 23 million cars were registered in the United States—one for every five Americans, up from one for every thirteen at the beginning of the decade. The next few decades were calmer in terms of growth, until sales again took off after the end of World War II. The number of vehicles in the United States alone was close to 50 million by 1950, approaching 75 million by 1960, and easily surpassing 100 million by 1970, when there was one car on the road for every two Americans.

By the 1960s, with automobile sales now also increasing markedly in Europe and Japan, governments and consumers began to pressure firms to produce new models with better environmental and safety features. Since then, there's been a steady, though inconsistent, improvement. Today's average automobile in North America, western Europe, and Japan is now much safer (with features like seat belts, antilock brakes, and air bags) and causes less environmental damage over its life cycle (with features like catalytic converters and fuel injection). The gradual declines in death rates per mile driven in every developed country and the cleaner air in local environments like California are clear evidence of these changes.

The benefits of these changes, however, have been highly unequal, accruing primarily to wealthy consumers in powerful states, and only secondarily to poorer consumers in more polluted environments. Moreover, the environmental impact of the global auto industry *overall* is rising steadily as more and more vehicles enter the roads. The number of passenger cars and commercial vehicles went from a handful at the end of the nineteenth century to well over 800 million today. And the numbers continue to rise.

By the middle of this century, assuming current trends hold, over 2 billion light motor vehicles will clog the world's roads. Most of this increase will occur in the developing world—places without the infrastructure or government capacity to maintain safe or healthy environments. Consumers will spend hundreds of billions of dollars to buy and run their cars and trucks. These direct costs, moreover, will be only a fraction of total costs, which also include outlays for roadwork, parking spaces, bridges, traffic cops, state regulation, and environmental cleanups. So many automobiles will also mean dirtier cities in the Third World (even as some cities in the First World become cleaner) and much greater stress on the global environment—adding considerably, for example, to global greenhouse gases. It will mean many more traffic deaths, too. Already the risk of dying is high, with lifetime odds of about 1 percent as a cause of death even in the wealthiest countries. Traffic collisions now kill over 1 million people a year—and injure as many as 50 times that number. Already 85 percent of these deaths are in developing countries, and the future there looks even bleaker. By 2020, the number of annual traffic deaths is expected to exceed 2 million, even though mortality rates are likely to continue to decline in all of the wealthy countries of the world.[1]

The shadow effects of the auto industry, then, are intensifying for poor people and the global environment, even as new technologies decrease per unit impacts of new vehicles. Rising sales of sport-utility vehicles over the last decade show the dangers of relying on technological advances, corporate interests, and consumer sovereignty to transform the societal and environmental impacts of this industry. Many consumers clearly enjoy the sense of power of these large, easy-to-handle vehicles. Many consumers *feel* safer, too (even though the statistics on occupant death per SUV on the road are not notably better). Yet the choice to drive an SUV increases the risks of injury or death for pedestrians as well as other drivers and passengers. These vehicles transfer ecological costs to others as well, occupying more space than cars and requiring more resources—such as gasoline and rubber and steel—to operate. They also strain the global environment more than cars, from stirring up desert dust storms to emitting more greenhouse gases. Already they account for nearly half of total vehicle purchases in the United States; and, if global trends continue, half of all motor vehicles worldwide will be SUVs or other light trucks by 2030.

Thus improvements increasing the safety and reducing the ecological impacts of automobiles are occurring too slowly to prevent irreversible harm both to millions of families (especially in developing countries) and to the global environment. In some ways, this outcome is a natural consequence of globalizing economic change itself, which is not primarily about creating better living conditions, but rather about generating more sales and profits. The current philosophy of Toyota—the Toyota Way— is revealing here. The *Economist* describes it as the "relentless pursuit of excellence."[2] So impressive is Toyota's record of steadily manufacturing more reliable, safer, and more environmentally friendly cars and trucks with less waste and fewer resources that, for decades, other auto firms have been borrowing Toyota's ideas—like just-in-time assembly lines—to improve quality and efficiency of production.

Yet this relentless pursuit of excellence has also meant a relentless pursuit of global sales. Toyota is already in the Global Ten—accounting for over 10 percent of annual vehicle production—and, by 2008, was right on the heels of the global leader, General Motors, with the aim of becoming the only auto firm in what it calls, in an effort to motivate workers, the "Global Fifteen." Toyota's tactics to increase its market share may differ somewhat from Ford and General Motors during the first half of the twentieth century. But the outcome is basically the same: more structural dependencies—personal, socioeconomic, and political—on more and more automobiles.

Some people are resisting the growing global dependency on the automobile. Citizens have mounted protests—some even successful—to block road construction. Some individuals choose to live without a car as part of a simpler lifestyle; some also choose to "jam" elite cultures with messages about the perils of driving. Others, such as Mothers Against Drunk Driving (MADD), are encouraging more responsible use of automobiles. Some communities like Singapore and London are charging drivers for access to downtown areas to reduce congestion. Others like Canada's Toronto Islands are banning automobiles outright.[3] A hornet's nest of critics is now attacking the rising numbers of SUVs. One campaign in the United States to encourage Americans to drive less (especially SUVs) is asking: "What would Jesus drive?"[4] Another links SUVs to oil imports and thus to funding Middle East terrorists. For all that, resistance is sporadic and fragmented, doing little to slow the rising global consumption of automobiles.

The next case tells the story of fueling the automobile with leaded gasoline. It begins nearly a century ago, with a team of scientists in a DuPont laboratory searching for a way to improve the performance of gasoline.

II

Leaded Gasoline

7

Leaded Science: Pumping Out Profits and Risks

The history of the ecological shadow of leaded gasoline begins on a Friday late in 1921. For five long years, Thomas Midgley Jr. and his team in the Fuel Division of the General Motors Research Corporation had been searching for an effective additive to increase the octane of gasoline, and thus reduce engine "knock" or pinging. Their successful test of tetraethyl lead at the DuPont plant on Friday, 9 December 1921, set DuPont and GM in motion. The new additive was effective, it was inexpensive, and a patent was easy to obtain.

Then, in October 1924, just as consumers all across the United States were beginning to switch to the "improved" gasoline, tragedy struck at several lead-processing plants. Before long, medical specialists and government regulators were questioning the safety of putting a known poison into gasoline (and thus also into automobile exhaust). The history of leaded gasoline could have ended—but did not—with this seemingly commonsense questioning. Instead, after a brief pause for "research," no one would again seriously question the safety of leaded gasoline until the 1960s, when geochemist Clair Patterson began to wonder why his tests were indicating such high lead levels in the northern hemisphere.

Why did U.S. regulators declare leaded gasoline "safe" in the 1920s? Why were critics, once so loudly calling for precaution and further research, so quiet in the following decades? And, finally, why were so many scientists, activists, and government agencies in the United States suddenly—a half century later—beginning to challenge the "facts" and "safety" of leaded gasoline?

The answers reveal important lessons about how and why ecological shadows form, intensify, and shift. They reveal how the pursuit of profits and markets can trump calls for precaution even in the face of high uncertainty and significant risks, how corporations and state allies can silence critics and control "research" for generations, and, finally, how knowledge and technology can both cause and begin to mitigate ecological shadows.

The Profitable Breakthrough

Knocking occurs when part of the fuel in a cylinder ignites prematurely. This makes a pinging sound, causes the engine to run rough, and eventually damages it. It was common in internal combustion engines before the 1920s, worsening with higher compression ratios and increased engine power. Midgley's team ran thousands of tests before discovering the value of tetraethyl lead as a fuel additive.[1] At first, the team was willing to test just about any compound—even melted butter. A few were near successes. Back in December 1916, Midgley found that iodine could eliminate knock, but, besides being highly corrosive to engine parts, it was expensive, adding a dollar to the cost of a gallon of gas. Two years later, Midgley and his team were encouraged to discover that aniline was even more effective than iodine, although still unsuitable.

The need for an effective antiknock agent was even greater after World War I, when the world turned to lower-quality oil reserves. By then, some geologists were estimating that reserves in the United States would run dry in a generation or two, and U.S. automakers were becoming increasingly eager to develop more efficient fuels.[2] Midgley and project head Charles F. Kettering did consider other options besides finding a low-cost and low-percentage antiknock additive for gasoline. One possibility was to replace gasoline with industrial (ethyl) alcohol from crops like corn and sugarcane. Another was to mix industrial alcohol and gasoline to create an antiknock fuel.

The team eventually dismissed both options on economic grounds. The main reason was straightforward: it was impossible to patent industrial alcohol. At the time, farmers could make it in a backyard still. Thus the potential for future corporate profits was far greater with a patented antiknock additive for gasoline. There were other reasons as well. The oil firms were opposing any shift away from gasoline. And there were practical barriers to a switch to industrial alcohol. "The present total production of industrial alcohol amounts to less than four percent of fuel demands," Kettering would later explain in a 1921 speech, "and were it to take the place of gasoline, over half of the total farm area of the United States would be needed to grow the vegetable matter from which to produce this alcohol."[3]

The search for a cheap and effective antiknock additive for gasoline was therefore continuing with ever greater intensity by the early 1920s. In the spring of 1921, Midgley's team found yet another chemical—the solvent selenium oxychloride—able to prevent knocking. Unfortunately, "it had some disadvantages," Midgley would wryly explain afterward, "not the least of which was its tendency to turn the engine into a chemical solution." Midgley's team doggedly went on testing more compounds. Soon they found that diethyl telluride was even more effective at eliminating engine knock, but it stank horridly, having, in Midgley's typically playful words, "a satanified garlic odor" that "would cling to you for weeks."[4]

These near successes helped Midgley narrow his search. Using the periodic table, he worked his way toward the heavy end of the carbon group, eventually reaching lead. This he tested in a compound called "tetraethyl lead," first discovered by a German chemist in 1854 but never used commercially because of its toxicity. The breakthrough came in the DuPont plant in a simple experiment in December 1921, when Midgley's team started up a one-cylinder Delco-Light engine filled with kerosene containing 0.025 percent tetraethyl lead. The engine ran smoothly—even better than with the adopted standard of 1.3 percent of aniline. Soon it became clear that a "spoonful" of tetraethyl lead, costing only a penny, "was enough to convert a gallon of gasoline from a rattling, knocking nuisance into a smooth-running motor fuel."[5]

The team quickly filed a patent for the blend of gasoline and tetraethyl lead, then turned to refining, manufacturing, and marketing the new fuel. They added ethylene dibromide to minimize the buildup of lead deposits in the engine (which damages spark plugs and valves). In October 1922, GM contracted DuPont to supply tetraethyl lead: Pierre du Pont signed as president of GM, while his younger brother, Irénée, signed as head of DuPont.[6]

Two months later, the U.S. Surgeon General wrote the chairman of the board of DuPont to ask whether the company was sure that tetraethyl lead was safe. This "had been given very serious consideration," Midgley wrote back on the chairman's behalf. He admitted that he did not have "actual experimental data." Still, "the average street," he wrote confidently, "will probably be so free from lead that it will be impossible to detect it or its absorption."[7]

The new fuel blend went on sale in February 1923 without any research on the potential health effects. Lagging well behind Ford, whose Model T sales were in full surge, General Motors was eager to introduce a gasoline that could eliminate engine knock: marketing its more powerful cars—such as GM's flagship Cadillac—was central to its comeback strategy. GM advertised its "ethyl gas" as "a better fuel for motors."[8]

Ethyl Gas Goes Loony

The General Motors Chemical Company was formed in the spring of 1923 to market ethyl gas: Kettering was president and Midgley vice president. The many advantages of the gasoline brought some easy marketing victories. In May of that year, the top three drivers at the Indianapolis 500 powered across the finish line on ethyl gas. As a sign of the growing importance of ethyl gas, General Motors and Standard Oil of New Jersey (which held a patent on a cheaper way to synthesize tetraethyl lead) formed the Ethyl Gasoline Corporation— renamed the Ethyl Corporation in 1942—in August 1924.[9] Kettering was again president; Midgley, now second vice president and general manager.

Not everything was proceeding smoothly, however. Keeping conditions safe for workers handling commercial volumes of tetraethyl lead was proving difficult. Even Midgley, aware of the risks of prolonged exposure to the additive from the outset, was not immune: he needed to spend a month in Florida in early 1923 to recover. There was a growing worry as well about the possibility of lead poisoning from the exhaust of automobiles running on ethyl gas. A few months after Midgley's respite in Florida, General Motors contracted the U.S. Bureau of Mines to investigate the health effects of tetraethyl lead.

Then, within a single week in October 1924, five workers died at the Bayway tetraethyl lead laboratory at the Standard Oil refinery in Elizabeth, New Jersey. Black-and-blue from muscle spasms, writhing in agony, increasingly delusional and paranoid, the poisoned workers became violent and suicidal. Doctors saw little choice but to restrain the most violent in straitjackets. Confusion and denial reigned at the Bayway facility. A Bayway manager in charge of some of the men, after learning of the death of one of his men and the grave illnesses of four others, thought for a bit, then jotted down an explanation as odd as the disease itself: "These men probably went insane because they worked too hard."[10] When reports of these deaths hit the front pages of newspapers across the United States, Midgley was dispatched to reassure the public.

Earlier in the year, after witnessing firsthand the hazards of its commercial production, Midgley had considered abandoning tetraethyl lead as an antiknock agent. But, by the time of the deaths at Bayway, he was again enthusiastic, perhaps because of advance notice that the Bureau of Mines would soon conclude that the exhaust from ethyl gasoline was safe. Midgley was a natural showman. To demonstrate that tetraethyl lead was harmless in small doses, he rubbed some into his hands during a press conference at the offices of the Standard Oil Company on 30 October 1924.[11] This stunt did little, however, to reassure worried state officials. That same day, the New York City Board of Health banned ethyl gasoline; the state of New Jersey and the cities of Philadelphia and Pittsburg soon followed. Sales went ahead elsewhere, although many authorities were now less than enthusiastic.

With the release of the Bureau of Mines' report the day after Midgley's press conference, ethyl gas received some positive news coverage. The *New York Times* headline on 1 November summed up the report's findings: "No Peril to Public Seen in Ethyl Gas, Bureau of Mines Reports after Long Experiments with Motor Exhausts." The methodology for this conclusion would later undergo intense scrutiny from other health researchers. Every day for 3 to 6 hours, bureau scientists had exposed over 100 types of animals, including monkeys, dogs, rabbits, and pigs, to leaded gasoline exhaust. After eight months of testing, the animals had shown no signs of lead poisoning (such as paralysis or loss of appetite and weight). "The danger of sufficient lead accumulation in the streets through the discharging of scale from automobile motors," the bureau concluded, was "seemingly remote."[12]

Still, with news continuing to emerge of more deaths of workers at the other two plants producing tetraethyl lead, Ethyl was now struggling to stay in business.[13] The Bayway plant remained closed following the deaths in October. The media were now calling ethyl gasoline the "loony gas." Workers at the DuPont plant in Deep Water, New Jersey, which first began producing tetraethyl lead in September 1923, were now calling their plant the "House of the Butterflies" because so many of them were hallucinating about winged insects, an early sign of lead poisoning. "The victim," explained *New York Times* reporter Silas Bent in 1925, "pauses, perhaps while at work or in a rational conversation, gazes intently into space, and snatches at something not there." Managers claimed workers in Deep Water were told to wear gloves, protective clothing, and gas masks. Still, over three-quarters were poisoned, some repeatedly, during the first year and a half of operations.[14]

Reassuring Consumers

The company took several steps to gain the public's confidence in the face of these grisly tragedies. Ethyl, along with Standard Oil and DuPont, convinced the Surgeon General in December 1924 to request a public airing of the health effects of tetraethyl lead. Then, during the resulting conference held by the Public Health Service in May 1925, Ethyl announced it would suspend sales to allow for further studies.[15] This conference gave Ethyl, Standard Oil, General Motors, and DuPont a public forum to make a case for the value of adding tetraethyl lead to gasoline. Frank Howard, a senior executive at Standard Oil, called tetraethyl lead a "gift of God" that allows "a gallon of gasoline . . . to go perhaps 50 percent further."[16] Many at the conference listened sympathetically to these arguments. Still, those attending heard enough counterarguments to decide to create a small committee under the Surgeon General to investigate the potential health effects of tetraethyl lead further.

The committee reported in January 1926. The deaths thus far, it decided, arose because of unsafe manufacturing practices: accidents that were avoidable with proper precautions. It found no hard evidence of poisoning from exposure to leaded gasoline, although it did urge the U.S. government to continue its research as "widespread" use of leaded gasoline could create "very different" conditions. The conclusion—"there are at present no good grounds for prohibiting the use of ethyl gasoline"— was a green light for the Ethyl Corporation.[17]

Ethyl gasoline went back on sale in May 1926. There were now stricter safety rules for manufacturing and distributing tetraethyl lead as well as signs posted at service stations along the lines of "Ethyl Gasoline containing tetraethyl lead, to be used as motor fuel only, and not for cleaning or any other purpose."[18] The company agreed as well to follow the voluntary standard set by the Surgeon General of a maximum of 3 cubic centimeters of tetraethyl lead per gallon of gasoline (equal to 3.17 grams per gallon). The Ethyl Gasoline Corporation began to advertise ethyl as an efficient gasoline sold by responsible oil companies able to enhance a car's power while reducing vibration (sometimes without even mentioning the lead additive). Kettering and Midgley were no longer in charge. Kettering had been demoted from president to director a year earlier. Midgley had gone from being second vice president and general manager of Ethyl to simply an employee of GM. "We felt that it was a great mistake to leave the management of the property in the hands of

Midgley," GM President Alfred P. Sloan would later explain, "who is entirely inexperienced in organization matters."[19]

In 2003, Thomas Midgley Jr. was inducted into the American National Inventors Hall of Fame in Akron, Ohio, for his discovery of ethyl gasoline. His entry credits him with enabling "airplane makers to develop more powerful engines, which gave the United States a decisive advantage during the Second World War. The increased engine horsepower also allowed for greater aircraft safety, reliability, and speed."[20] Adding lead to fuel no doubt enhanced the performance and efficiency of automobiles and airplanes. Chemist George Kauffman estimates that tetraethyl lead "saved the public . . . one-third of the total gasoline costs that would have been paid had [it] not been discovered."[21] Yet leaded gasoline has had consequences far graver than Midgley or the Bureau of Mines or the Surgeon General could ever have foreseen.

Silencing the Critics

Not all scientists agreed with the reassuring reports by the Bureau of Mines and Surgeon General in the 1920s. One of the most vocal critics was Yandell Henderson, professor of Applied Physiology at Yale University. In one speech in April 1925, for example, he damned the bureau's findings, saying "the investigators in the Bureau of Mines have used experimental conditions which are fundamentally unsuited to afford information on the real issue." The approach to studying the effects of acute lead poisoning may have been reasonable. But, in his view, the research on the potential long-term health effects of lead emissions from automobile exhaust was, quite simply, shoddy and misleading.

The reason, Henderson said, was straightforward: General Motors funded the bureau's research. The findings, he explained, were really about protecting "billions of dollars" in future profits for General Motors, Standard Oil, DuPont, and the Ethyl Gasoline Corporation, not about protecting public health. These were powerful financial interests. Given this, he could see little chance that the U.S. government would ban leaded gasoline unless it suddenly began to kill large numbers of people.

"It seems more likely," he said presciently, "that the conditions will grow worse so gradually and the development of lead poisoning will come on so insidiously (for this is the nature of the disease) that leaded gasoline will be in nearly universal use and large numbers of cars will have been sold that can run only on that fuel before the public and the

Government awaken to the situation." This would make it extremely difficult to eliminate lead from gasoline by the time authorities awoke to the dangers. Future control was, according to Henderson, the real reason for the enthusiastic corporate reaction to the discovery of tetraethyl lead. "It is sold now at little or no profit," he explained. "The profit is in the future and will consist in control of the gasoline business and control of the automobile industry. The power or combination of powers that holds the patents on tetraethyl lead will be the only one that will be able to make the cars that we all want to buy."[22]

A few other prominent professors and scientists were just as concerned over the use of ethyl gasoline on health grounds. David Edsall, dean of the Harvard School of Public Health, and a member of the 1925 Surgeon General's expert committee on tetraethyl lead, called the final report "half-baked." Alice Hamilton of Harvard University spoke repeatedly about the potential dangers of tetraethyl lead. Her views are perhaps best expressed by her comment to Kettering during a break in the May 1925 conference. Glaring at him, she blurted: "You are nothing but a murderer." There were critics outside of the United States as well. Erik Krause at Potsdam Institute of Technology in Germany, for example, wrote a letter to Midgley calling tetraethyl lead "a creeping and malicious poison," so deadly it killed a member of his dissertation committee.[23]

The counterattack against these critics was sharp. The editor of *Chemical and Metallurgical Engineering*, H. C. Parmelee, called them "incompetent," "misguided zealots," putting forth "hysterical testimony" that ignored that the research on tetraethyl lead was "in a fine spirit of industrial progress looking toward the conservation of gasoline and increased efficiency of internal combustion motors." The efforts to counter opponents of leaded gasoline became better coordinated after 1928 with the formation of the Lead Industries Association. In the end, critics like Professor Henderson did little to slow the sales of leaded gasoline, and in an impressively fast takeover of market share, 90 percent of all of the gasoline in the United States was leaded by the 1930s.[24]

The Decades of Industry Science

The U.S. government didn't follow the advice of the 1925 Surgeon General's committee to continue to investigate the long-term health and environmental effects of widespread use of leaded gasoline. Instead, over the next four decades, research on leaded gasoline was geared toward serving industry interests.[25] The most notable "industry scientist" was the toxicologist Robert Kehoe, Ethyl's chief medical consultant from 1924 until retiring in 1958. He was also the founding director of the Kettering Laboratory of Applied Physiology at the University of Cincinnati. Opening in 1930 with a donation from General Motors, DuPont, and Ethyl, the laboratory was the world's chief source of data on the health effects of leaded gasoline for over three decades.

The lab's findings were consistent: leaded gasoline was in no way a danger to public health. Three core assumptions about lead grounded this conclusion. It was natural to find some lead in human blood; lead was safe below a certain threshold; and the levels in countries like the United States were far below this threshold, meaning any lead emitted from automobile exhaust was, in effect, harmless. The U.S. government became so confident in this research it agreed to raise the voluntary standard from 3 to 4 cubic centimeters of tetraethyl lead per gallon of gasoline in 1958 (equal to 4.23 grams per gallon).[26] "During the past 11 years, during which the greatest expansion of tetraethyl lead has occurred," explained the Surgeon General at the time, "there has been no sign that the average individual in the United States has sustained any measurable increase in the concentration of lead in his blood or in the daily output of lead in his urine."[27]

The industry's control over research took its first hit in 1965 with Clair Patterson's groundbreaking analysis of the contaminated and natural levels of lead in humans in the northern hemisphere, which would be published in the September issue of *Archives of Environmental Health*. Patterson, a geochemist at the California Institute of Technology, first wondered about the extent of lead contamination during his work on dating the age of the earth more accurately at 4.55 billion years. The estimates in his 1965 article were alarming. Residents of the United States had concentrations of lead around 100 times higher than natural levels, while the level of lead in the atmosphere of the northern hemisphere was more than 1,000 times higher than natural levels. His recommendation was blunt: eliminate the main sources of lead pollution, including ethyl gasoline.[28]

According to Patterson, as soon as Ethyl learned of his findings—some months *before* they were published—it reacted by offering him "research support" to "yield results favorable to their cause" and, when he refused, by launching a campaign to discredit his results. In a review of the article for the editor of *Archives*, Robert Kehoe called it "remarkably naïve," the "brash" musings of someone who knew too much about rocks and next to nothing about toxicology. Still, he would "welcome" the publication of the article, so the findings, which he felt were spreading by "word of mouth," could be "faced and demolished."[29] Guided by Kehoe, the orthodox establishment was soon ridiculing Patterson's research. Efforts, too, were made to block any further research by him on lead. The Public Health Service and the American Petroleum Institute cut off his research money. And pressure was put on Cal Tech to dismiss him.

A 1966 hearing of the Senate Subcommittee on Air and Water Pollution was a turning point for both Kehoe and Patterson. In the wake of layoffs and falling revenues, Ethyl was lobbying the U.S. Public Health Service to raise the amount of lead permitted in a gallon of gasoline. Kehoe was the star expert witness for Ethyl. He had, as he explained with pride to the chair of the subcommittee, "more experience in this field than anyone else alive." The chair was obviously wary of the arrogance of Kehoe's certainty and prodded him to explain why others, including the Public Health Service, were starting to question the accuracy of his conclusions.

Still, Kehoe didn't waver. There was, he said, "not the slightest evidence" of harm from airborne lead, and leaded gasoline posed no risk at all to public health. But when Patterson explained his research findings to the hearing a week later, he could see little reason for Kehoe's complacency. The public health threat from lead, he argued, was more appropriately seen on a continuum, with acute poisoning at one end and chronic low-dose poisoning at the other.[30] With some in the U.S. government now listening to Patterson, other researchers were also beginning to investigate the effects of exposure to low doses of lead, especially on vulnerable groups such as children and pregnant women. After 40 years of industry control over the research on the health effects of leaded gasoline, the policy and science communities were beginning to turn against Kehoe and the Kettering Laboratory.

The Tide Turns

By 1970, the U.S. Surgeon General was calling for research to identify children with "undue" lead exposure. For the first time, the federal government also began to fund studies of the health consequences of lead for children. "The industrial monopoly on scientific data," child psychiatrist Herbert Needleman wrote about this period, "was drawing to an end."[31]

The consensus within the industry over the value of putting lead into gasoline was breaking down as well. General Motors had sold the Ethyl Gasoline Corporation to Albemarle Paper in 1962, which, being 18 times smaller than Ethyl, had to borrow $200 million to acquire it.[32] Although the reasons for the sale are not fully clear, Ethyl's flat profit line at the time was certainly a factor (Ethyl's patent on tetraethyl lead expired in 1947), as were worries about possible antitrust investigations.

This sale left Ethyl a much weaker player in the corporate lobby of the U.S. Congress. Still, Ethyl executives felt shocked and betrayed when GM announced in January 1970 that it would install catalytic converters by 1974. Because lead destroys the platinum catalyst in the converter, cars with catalytic converters need unleaded gasoline. Later, GM would argue it had little choice, with the 1970 Clean Air Act about to set mandatory targets to decrease automobile emissions, including 90 percent reductions in carbon monoxide, nitrogen oxides, and hydrocarbons. The official biographer for Ethyl saw it differently. "It struck some people as incongruous—not to use a harsher word—for General Motors to sell half of what was essentially a lead additive firm for many millions and then advocate annihilation of the lead antiknock business."[33]

Before long, other automakers were announcing similar plans to install catalytic converters. Suddenly, the future for the lead additive industry was looking grim. The chairman of the Mobil Oil Corporation, in a speech in January 1970, was blunt when explaining the thinking among automakers: "Lead must go."[34]

8

Lead Must Go

The leaded gasoline industry in the United States didn't simply roll over and concede that "Lead must go." On the contrary, in the early 1970s, the industry association embarked on a full-scale campaign to discredit critics, block legislation, and delay environmental regulations—it put ads in newspapers, lobbied politicians and bureaucrats, sued the Environmental Protection Agency, and worked hard to keep all of the corporate players on board. Nevertheless, as domestic sales of automobiles with catalytic converters (which worked only with unleaded gasoline) rose, support to phase down leaded gasoline grew, even among U.S. automakers and major oil companies.

Cornered, the U.S. leaded gasoline industry fought to survive, pursuing multiple strategies to maintain markets and profits. But to no avail. Despite its best efforts to stall, if not avert, a phasedown, by the mid-1980s the industry was effectively out of business, at least within the United States. The result was a much lighter ecological shadow of lead over the United States—and a much healthier environment for Americans—with average blood-lead levels dropping sharply by the late 1980s.

Why did this occur? The answer reveals further lessons about how and why ecological shadows of consumption change. It shows how important the strengthening of environmentalism since the 1960s was for changing regulatory frameworks and consumer choices. And it shows how introducing new environmental technologies—in this case, the catalytic converter—can alter trade and markets in ways that weaken an industry's capacity to maintain political and corporate allies in the face of environmental criticism. At first glance, this might seem to suggest that environmentalism, technological ingenuity, and scientific knowledge enjoyed unqualified success in overcoming corporate resistance. Indeed, governments deciding to follow in the footsteps of the United States were often able to impose phasedowns more quickly.

Yet the phasedown of leaded gasoline in the United States, along with phasedowns in other (mostly wealthy) countries, would end up intensifying the ecological shadows of leaded gasoline in many of the world's poorest regions and most unstable ecosystems. The chapter begins with an analysis of corporate efforts in the first half of the 1970s to delay any phasedown in the United States—essential for understanding the eventual results, and compromises, of the following decade.

Delaying the U.S. Phasedown

By 1970, little headway had been made to reduce the consumption of leaded gasoline in the United States: 98 percent of gasoline still contained lead, and automotive exhaust accounted for around 80 percent of airborne lead.[1]

The 1970 Clean Air Act directed the Environmental Protection Agency to set standards for safe levels of exposure to pollutants. Although, at the time, the agency didn't have much experience with lead, the first EPA administrator, William Ruckelshaus, seemed ready to tackle the task quickly, stating in 1971 that tetraethyl lead was "a threat to public health."[2] Predictably, scientists with links to the lead industry disagreed vehemently and, in an about-face, were now arguing that extrapolating from tests on animals to humans was of questionable value. The Ethyl Corporation was also claiming that the levels of lead from ethyl gasoline were inconsequential compared with those from lead-based paint. Some within the EPA became unsure about the need to regulate leaded gasoline, and a few within the agency were even arguing against doing so.

The agency decided to contract the National Academy of Sciences to survey airborne lead and provide advice. The resulting 1972 report was vague and cautious, hardly surprising given that the team of consultants included scientists like Robert Kehoe, but not like Clair Patterson. The team found no conclusive data to indicate levels of atmospheric lead below 2 micrograms per cubic meter raised average blood lead levels. Nor did it find any evidence that low levels of lead were toxic. A senior review committee for the National Academy of Sciences was not impressed. The consultants "failed miserably," the chair of a review team lamented, "to form any sort of precise conclusion."[3] But their report did provide an excuse for delaying a phasedown to allow more time for research, an outcome that pleased Ethyl executives and investors (the corporation's stock jumped 20 percent a day after the report was released).

Over the next year, the EPA became increasingly concerned with the health consequences of chronic exposure to the 180,000 metric tons of lead spewing from vehicle exhaust every year. The November 1973 agency report, *EPA's Position on the Health Implications of Airborne Lead*, was firm in its conclusion: lead from automobile exhaust was an immediate threat to public health.[4] The EPA knew the decision by automakers like GM to introduce catalytic converters would soon begin to chip away at demand for leaded gasoline. But it was worried automakers would find a way to modify catalytic converters to function with leaded gasoline. A month after the November report, the EPA announced a gradual phasedown, starting 1 January 1975, to reduce the average amount of lead in a refinery's total gasoline output: from an average of 2.2 grams per gallon to 1.7 grams in 1975, then down to 0.5 grams per gallon by 1979.

These EPA targets were not met. DuPont and the Ethyl Corporation quickly sued. With a touch of Orwellian irony, Ethyl called its legal team the "Ethyl Air Conservation Group." The firms won the first round in late 1974. A panel of the U.S. Court of Appeals for the District of Columbia set aside the EPA's standards, calling them "arbitrary and capricious" and agreeing with Ethyl's argument that the EPA must demonstrate "actual harm" rather than just "significant risk."[5]

Legal wrangling was not the only reason for delays in regulating a phasedown of leaded gasoline. Political struggles were occurring both within the EPA and between the EPA and other government agencies. The public affairs branches of the lead additive industries were busy as well, portraying the efforts to remove lead from gasoline as a dangerous waste in a time of growing oil shortages and price hikes. A full-page newspaper ad running from late 1973 to early 1974 was typical. It showed oil pouring down a manhole from a drum draped in an American flag. The caption was designed to unnerve: removing lead from gasoline would waste one million barrels of American oil every day.[6]

Still, the EPA was putting in place policies that would bring about a significant phasedown of leaded gasoline. The agency made catalytic converters and special, narrower fuel inlets mandatory beginning with 1975 models. It required large retailers of unleaded gasoline to design nozzles to fit the narrower fuel inlets (which the standard, wider fuel nozzles for leaded gasoline did *not* fit). Refiners failed to convince the courts to overturn these measures and, beginning in 1975, unleaded gasoline pumps began to appear across the United States to serve cars with catalytic converters.

The EPA won a significant victory in 1976 when the full U.S. Court of Appeals for the District of Columbia overturned the 1974 decision and reinstated the lead standards, ruling in a 5–4 majority that the EPA could act on the basis of significant risk. The statute, the court advised, is "precautionary in nature and does not require proof of actual harm before regulation is appropriate."[7] The firms tried to fight on with an appeal, but the Supreme Court declined to hear it.

The U.S. Phasedown Accelerates

After these court decisions, the phasedown began to produce significant results. The amount of lead consumed in gasoline from 1976 to 1980 fell by half; the concentrations of airborne lead fell immediately. In Philadelphia, for example, the range was 1.3 to 1.6 micrograms per cubic meter in 1977. Just three years later, it was 0.3 to 0.4 micrograms per cubic meter—less than one-quarter as much. Over this time, similar declines occurred across all major cities in the United States, with average levels of lead in human blood falling 37 percent.[8]

The growth of self-service gas stations in the late 1970s presented a minor setback for efforts to rid U.S. roads of leaded gasoline. Some drivers of cars with catalytic converters—the EPA put the total at 10–15 percent—switched back to leaded gasoline. Some would take along a funnel when they refueled; others would make the fuel inlet large enough to fit a leaded gasoline nozzle. Most of these "fuel switchers," it seems, were trying to save money: unleaded gasoline in the late 1970s and early 1980s cost 6–14 cents per gallon more than leaded gasoline. It was illegal for attendants and commercial fleet owners to refuel cars with catalytic converters with leaded gasoline, but not for the average driver. Owners faced a penalty for failing a vehicle emissions inspection, but many of the inspection programs were unable to detect a destroyed catalytic converter.

The compliance rate was nevertheless reasonable. Most drivers simply obeyed the label, "unleaded gasoline only." Others seem to have been genuinely worried about the long-term health effects of using leaded gasoline for the general public, and especially for children, the subject of some 150 articles from 1970 to 1990, many in popular magazines like *Newsweek* and *Time*. Even by the early 1970s, pollsters were finding that between 60 and 70 percent of Americans worried about leaded gasoline exhaust "often" or "sometimes."[9]

The U.S. Lead Industries Fight Back

Far from conceding defeat, the Lead Industries Association petitioned the EPA in 1980 to revoke the new standard for ambient lead levels. Its case seemed strong, at least on its face. The EPA had justified the standard with an Idaho Health Department study estimating that a 1 microgram per cubic meter increase in atmospheric lead raises blood-lead levels by 2 micrograms per deciliter (2 millionths of a gram per 100 milliliters). One of the coauthors, Anthony Yankel, was now saying he had miscalculated and overestimated the impact of atmospheric lead. The judge ruled against the Lead Industries Association, finding the original study to be sound; moreover, on learning that Yankel had gone to work for the lead industries after leaving the Idaho Health Department, he urged the Department of Justice to investigate Yankel's conduct.[10]

The election of Ronald Reagan as president gave the lead industries new allies. In 1981, the Reagan administration slashed the EPA's budget, personnel, and powers to enforce regulations. At first, it seemed the EPA would rescind its lead standards. But, with mounting scientific evidence of the toxic health effects of low levels of lead on children, and after a public backlash and much politicking between various agencies and interests, the Reagan administration realized that the political consensus was now strongly in favor of eliminating leaded gasoline and backed off.

The result in 1982 was a tightening of the phasedown rules for small refineries. The EPA also recalculated limits for lead content. It was now an average of the amount of lead in leaded gasoline alone, rather than the amount of lead in the entire gasoline pool. This was significant because unleaded gasoline was taking up an increasingly large portion of total gasoline sales. The new rules required refineries to maintain a quarterly average of 1.1 grams or less of lead per gallon of leaded gasoline (with some exceptions for very small refineries). This was not much different than the 1980 standard of 0.5 grams per gallon of total gasoline. But it was nevertheless a critical first step on the path to much bigger decreases in the near future.[11]

The public rhetoric of corporate executives over the next few years echoed that of the 1920s. The comments by Ethyl's Lawrence Blanchard in the early 1980s are typical. "It was misleading at best and fraudulent at worst to talk about symptoms and horrors of lead poisoning," he said at an EPA hearing. "That is just like talking about the horrors of gassing World War I soldiers with chlorine at a hearing as to whether we should chlorinate to purify drinking water."[12] By this time, however, such comments were winning over few supporters. Indeed, by 1984, the representatives and scientists working for the lead industries realized the battle to block the U.S. phasedown was all but lost. "Unfortunately, the atmosphere we're now in prohibits objective scientists from coming forward," lamented Ethyl's director of air conservation in a 1984 *New York Times* article. "And why should they, when they would be crucified by the press, the EPA and the environmentalists?" By then, some refiners were supporting stricter standards—such as Ashland Oil during the 1984 Senate hearings—in part because these firms had put in place the facilities to manufacture unleaded gasoline.[13]

In their ongoing efforts to delay the phasedown for as long as possible, the lead industries continued to harass leading researchers, such as psychiatrist Herbert Needleman, who had caused a commotion in the 1970s with studies correlating levels of lead among children with lower performance in school. A scientist with links to industry accused Needleman in 1982 of manipulating his data to show that lead was toxic, a charge the EPA science advisory council would later dismiss as baseless.[14] Charges such as "scientific misconduct," even when totally false, were a valuable tactic the lead industry used to taint the popular press with uncertainty, cause publishing delays, rattle senior scientists, and intimidate postdoctoral fellows and tenure-tracking assistant professors.

Still, by the mid-1980s, these firms knew their cause was hopeless. The research pointing to the unhealthy effects of leaded gasoline that started with Patterson's exploratory 1965 article was now an avalanche of hard-hitting data.

The U.S. Phaseout

In 1985, 40 percent of the gasoline sold in the United States was still leaded. That year, riding a wave of research showing the harmful effects of airborne lead, the EPA lowered the refinery pool standard to 0.5 gram of lead per gallon of leaded gasoline. The next year, the EPA dropped it to 0.1 gram. The "phasedown" was fast becoming a "phaseout."

In addition to setting tougher standards, the EPA employed a mix of policies to encourage refiners to cooperate and comply. The regulations from 1982 until 1987, for example, allowed refineries to trade lead permits through a system of averaging across the various refineries. The rules from 1985 to 1987 further permitted the banking of lead credits.

This gave the refinery industry as a whole more flexibility. In particular, it helped smaller refineries manage some of the compliance costs of removing lead and installing processing equipment to add lost octane. The trading and banking programs ended in 1988, when the EPA imposed a standard of 0.1 gram of lead per gallon of gasoline for individual refineries.[15]

The results of the U.S. phasedown from the mid-1970s to the late-1980s are impressive. The lead content of gasoline decreased by 99.8 percent from 1976 to 1990, and average blood-lead levels fell sharply as a result. The Department of Health and Human Services, surveying 60 cities across the United States, found a 78 percent drop in human blood-lead levels from 1978 to 1991. Today the U.S. medical threshold for "safe" blood-lead levels is below 10 micrograms per deciliter of blood (10 millionths of a gram per 100 milliliters).[16] Nearly 78 percent of children between 1 and 5 years old still had lead levels above 10 micrograms per deciliter of blood for the period 1976–80, whereas only 4.4 percent had such levels by the period 1991–94. The regulations to remove lead from gasoline, Needleman estimates, "spared as many as 3.4 million children from growing up with hazardous concentrations of the toxic metal in their bodies."[17]

The United States has only allowed small amounts of leaded gasoline since the late 1980s. The total amount of lead in gasoline was a little over 180,000 metric tons in the mid-1970s; by 1990, the total was around 450 metric tons—a decrease of 99.75 percent. Leaded gasoline accounted for a mere 0.6 percent of total gasoline sales when the EPA put in place a full ban for all highway vehicles in 1995; today only farm vehicles, marine engines, and racing cars are allowed to use leaded fuel.

A few countries, such as Japan, phased out leaded gasoline before the United States.[18] But most did not. Understanding when, how, and why other countries, especially those in the developing world, began to consume—and then phase out—leaded gasoline reveals the complex political and economic forces shaping the global consumption of gasoline.

Exporting Shadows of Lead

Leaded gasoline first began to spread beyond the United States in the 1930s. The Ethyl Gasoline Corporation set up Ethyl Export in the United Kingdom in 1930 to sell leaded gasoline overseas. This company became the Associated Ethyl Company in 1938, before changing its name to the Associated Octel Company in 1961. It first began to produce leaded gasoline in 1954. Four years later, the Lead Industries Association and the former American Zinc Institute formed the International Lead Zinc Research Organization to promote the global spread of leaded gasoline.

Firms like Ethyl reacted to the phasedown of leaded gasoline in the United States by expanding markets overseas. With sales falling in the United States, Ethyl tried to reassure shareholders by trumpeting its profits from a tenfold expansion of overseas business between 1964 and 1981 as a way to finance diversification. By the second half of 1979, Ethyl had announced that, for the first time, foreign sales of "antiknock compounds" surpassed U.S. sales.[19]

The spread of leaded gasoline overseas was uneven, with different countries having different degrees of exposure at different points in time. One of the countries with highest exposure was Mexico in the early 1970s, when the average amount of lead in the gasoline pool (both leaded and unleaded) was about 4.0 grams per gallon. Others were Egypt in the mid-1980s, with 3.1 grams of lead per gallon of gasoline, and Sri Lanka in the early 1990s, with 3.2 grams per gallon.[20]

Governmental success in phasing out leaded gasoline was uneven as well. A few did so in a matter of months. Others took much longer. Still others are yet to begin. Globally, the amount of lead added to gasoline declined by 75 percent between 1970 and 1993.[21] The U.S. phaseout accounted for much of this decline. But other countries, such as Brazil, Canada, Germany, Italy, Japan, Malaysia, Poland, and South Korea, were all part of this global phasedown as well. Wealthy states tended to remove lead from gasoline before poorer ones. By the mid-1990s, Japan, Canada, and the United States had phased out leaded gasoline. Germany was close behind, with only 2 percent of its gasoline still containing lead. On the other hand, 90 percent of the gasoline in the Philippines and 99 percent in Indonesia still contained lead at this time.

Many factors besides national income can influence phaseouts, in particular, factors related to the politics and economics of the auto and refinery industries. A few developing countries, such as Brazil and Thailand, were no longer consuming leaded gasoline by the mid-1990s.[22] At the same time, a few advanced economies still allowed significant amounts of leaded gasoline. In Britain, for example, 32 percent of the gasoline sold contained lead. The figure was 55 percent in Australia, higher than either China (40 percent) or Mexico (44 percent). By the end of the 1990s, however, the pattern in the global consumption of leaded gasoline was unambiguous: 80 percent of the countries continuing to use leaded gasoline were poor.[23]

The "Worldwide" Phasedown

A glance at Europe shows the complexity of the global pattern of phasing out leaded gasoline. Although Germany began to phase out leaded gasoline in 1972, the majority of western European states didn't begin doing so until the mid-1980s, and the European Union didn't ban leaded gasoline until 2000. One reason it took Europe so much longer than the United States was the strong opposition of European automakers and many European states to making catalytic converters mandatory. In part, this was a reaction to the United States having done so in 1975: many Europeans worried a similar policy for the EU would give a competitive advantage to U.S. firms exporting automobiles to Europe. Still, the number of cars with catalytic converters grew steadily in Europe after the mid-1970s, with the result that once the phaseout of leaded gasoline did at last begin, it proceeded faster in many European states than it had in the United States.[24]

Unlike the United States, many states in western Europe used differential taxes to lower pump prices for unleaded gasoline, thus creating fewer incentives for drivers with catalytic converters to use funnels to fill up with leaded gasoline. This policy, explains economists Henrik Hammar and Åsa Löfgren, "did not assume people would make enlightened, 'green' choices."[25]

Many governments of states outside of Europe were also able to phase out leaded gasoline faster than the United States, often by taking what environmental economist Hank Hilton calls a "giant step"—imposing one very large reduction in a two-year period during the phaseout process. As in Europe, many seemed to learn how to accelerate the process by studying the histories of earlier phaseouts. Many were convinced of the urgent need to act by the research showing the damaging health effects of exposure to low levels of lead. Moreover, the close correlations between falling amounts of leaded gasoline and falling blood-lead levels in countries like the United States made it clear that a phaseout could produce immediate results. Many governments came to see that the economic benefits of phasing out lead outweighed the costs, which continued to fall after the mid-1980s, as the global markets for unleaded gasoline grew. Shifts in economic conditions and technological possibilities also created more opportunities for governments to collaborate with oil companies and automakers in their phaseouts.[26]

By the mid-1990s, the main producer of tetraethyl lead, Octel Associates, was drawing on its profits from selling tetraethyl lead to spend up to $200 million a year to diversify,[27] even as it did its best to delay for as long as possible a worldwide phaseout of leaded gasoline. Deploying tactics very much like the ones by Ethyl Corporation decades earlier, it challenged the scientific findings that lead was toxic. In New Zealand, for example, it ran ads claiming lead was "naturally occurring"—like salt, alcohol, and sugar. It funded new studies that "failed" to find a correlation between declining levels of leaded gasoline and declining blood-lead levels. And it added in a new twist. Refiners should not move too fast to eliminate lead, Octel argued, because the benzene they were using instead to boost the octane of unleaded gas was a known carcinogen. "We're trying," explained a member of Octel's management board without a hint of irony, "to bring reason to the debate."[28]

But, as in the United States in the 1980s, by the late 1990s, it became clear Octel was waging a hopeless campaign. Today less than 10 percent of gasoline worldwide contains lead—down from 40 percent in 1991. Most of the remaining countries with leaded gasoline have developing economies, and even they are changing over to unleaded gasoline. As latecomers, they can take advantage of greater opportunities to phase out leaded gasoline even faster than before, as the case of sub-Saharan Africa will make plain in chapter 9.

9

Taking the Lead Out of Africa

Over the last decade, many developing countries have managed to phase out leaded gasoline. Many of these phaseouts have been easier and faster than in places like the United States. Why has this been possible? Answering this question helps explain why ecological shadows of consumption can—and do—sometimes quickly fade away, even in the world's poorest places. To that end, this chapter analyzes the phaseout of leaded gasoline in sub-Saharan Africa from 2002 to 2006. Broadly, it shows how global environmentalism—combined with international aid, corporate interests, and local political will—was able to accelerate environmentally friendly change in the region.

Although this success story in no way excuses the hypocrisy of corporations or states flooding this region with leaded gas, even after all were well aware of the dangers, it does reveal how actors and forces that once deflected ecological costs onto poor people and into weak states can later act to reduce and even eliminate those costs by helping alter local consumption to improve living conditions. (Chapter 24 will return to this ground for optimism when discussing possible strategies to reform the global political economy.)

The phaseout of leaded gasoline in sub-Saharan Africa took place in record time. Many factors explain why this was so. Knowledge and experience inspired some local governments and consumers to act. Global institutions such as the United Nations Environment Programme and the Global Environment Facility, lending organizations such as the World Bank, international meetings such as the 2002 World Summit on Sustainable Development, and national agencies such as the U.S. Environmental Protection Agency helped both to provide much-needed financial assistance and to maintain a scientific consensus to act. But, as this chapter will show, the principal reason was the shift among the most powerful corporations and states away from leaded and toward unleaded gasoline (and the automobiles that run on it).

Partnering for a Phaseout in Sub-Saharan Africa

At the conclusion of a regional conference hosted by the United Nations Environment Programme (UNEP) in Senegal, the 2001 Dakar Declaration called for a phaseout of leaded gasoline in sub-Saharan Africa. Delegates were alarmed by studies in Africa showing that lead from leaded gasoline exhaust impaired brain function in children and increased the risk of health problems in adults. They agreed to aim for a full phaseout of leaded gasoline "as soon as possible," but no later than 2005. This was an ambitious timeline. At the time of the conference, Sudan was the only one of the 49 countries in the region to rely solely on unleaded gasoline. And Sudan's reasons were primarily economic. The year before the conference, the government had switched to unleaded gas after opening a new refinery able to produce unleaded gasoline for both export and the domestic market.[1]

The 2002 World Summit on Sustainable Development in Johannesburg spurred the regional phaseout process along. The Partnership for Clean Fuels and Vehicles, a nonbinding public-private initiative taken at this summit to reduce vehicular air pollution in developing countries, became the core of a collective effort to phase out leaded gasoline from the entire sub-Saharan region by the end of 2005. The partners—who would eventually number over 70, including African governments, NGOs, research institutes, oil companies, private donors, the UNEP, the WHO, the EPA, and the World Bank—agreed to support cleaner fuel standards and cleaner vehicles, with the UNEP also agreeing to act as a clearinghouse for collecting and exchanging information as well as creating and distributing "fact packs."

Implementing the Partnership

After Johannesburg, members of the partnership began to provide technical and policy advice to African governments. The partnership raised money to assist these governments ($500,000 was already pledged by March 2003), held conferences and workshops with the support of the World Bank and UNEP, and ran campaigns to raise awareness in sub-Saharan Africa about the health consequences of leaded gasoline. The research findings were disturbing. According to Robert De Jong (a program officer at the UNEP), one study found that exposure to lead from leaded gasoline exhaust was lowering the IQ scores of children in major cities in Africa by 4 to 5 points.[2]

Progress was swift after Johannesburg. By 2004, seven more countries in the region—Cape Verde, Eritrea, Ethiopia, Mauritania, Mauritius, Nigeria, and Rwanda—had phased out leaded gasoline. Some, such as Ethiopia in January 2004, simply banned the import of leaded gasoline with little public debate or knowledge. This forced consumers to use unleaded gasoline for older cars without catalytic converters (which, at the time, constituted the bulk of cars across Africa). Other countries, such as Mauritius in 2002, first launched a campaign to educate the public about the benefits of a phaseout (including charts at filling stations to help consumers choose the best gas for older vehicles) before imposing a full ban on leaded gasoline. International agencies also worked to convince consumers to use unleaded gasoline.[3]

Phasing out was harder in countries where consumers could choose between leaded and unleaded gasoline. In Kenya, for example, many consumers continued to use leaded gasoline, believing it made older cars run better. This was a significant cause of air pollution. Thus, according to tests in 2005, emissions from an average car in Nairobi were 16 times greater than from an average new car in the United States, even though 70 percent of the automobiles in the capital had catalytic converters. A key reason was leaded gasoline, which had ruined most of those converters. Although the Kenyan government banned leaded gas imports in 2004 when imports accounted for some 30 percent of the country's total gasoline pool, the main source of domestic gasoline, a 1960s refinery in Mombasa owned jointly by the government, Caltex Oil, Kenya Shell, and British Petroleum, could process only leaded gasoline.[4]

As of May 2004, the prospects for meeting the ambitious end-of-2005 deadline set by the Dakar Declaration did not look good: 40 countries in sub-Saharan Africa were still using leaded gasoline, and just under half of all gasoline in the region still contained lead. In Kenya, just 4 percent of motor fuel was unleaded. Yet, with the assistance of the Partnership for Clean Fuels, countries like Kenya and Senegal did manage to keep their promises to phase out leaded gasoline by the end of 2005. The Mombasa refinery in Kenya stopped producing leaded gasoline as of 1 December 2005 and was handling enough unleaded gasoline by early 2006 to meet domestic demand (with government support for further upgrades). Cameroon, which exported leaded gasoline to countries like the Central African Republic, Chad, and Equatorial Guinea, also kept its promise to stop by the end of 2005.[5]

Unleaded Africa

No country in sub-Saharan Africa was importing or refining leaded gasoline by the beginning of January 2006; 16 countries in the region, including South Africa, had stopped importing or refining leaded gasoline in the previous month. It was just 10 years earlier that South Africa first gave consumers the option of purchasing unleaded gasoline, at a time when the country had some of the highest levels of lead ever recorded in children. It cost refineries in South Africa $1.6 billion to convert their facilities to handle unleaded gasoline. Like many other countries in the region, South Africa also introduced lead replacement petrol (LRP), charging the same price as unleaded gasoline for the same octane, with additives to protect the valve seats of older vehicles designed to run on leaded fuel.[6] If the U.S. experience is any guide, decreases in blood-lead levels will occur rather quickly and will have many long-term societal benefits, especially for at least 600,000 children in South Africa whose blood-lead levels were above 10 micrograms per deciliter at the beginning of 2006.[7]

The overall result a half decade after the Dakar Declaration is impressive. The sub-Saharan region, despite suffering some of the world's worst poverty and social chaos, has become the first developing region to neither produce nor import leaded gasoline. Although it took many months to remove all of the leaded gasoline from circulation in the last countries to act—Mozambique, South Africa, Zambia, and Zimbabwe—it's reasonable to characterize 2006 as the year the gasoline in sub-Saharan Africa became virtually lead free.[8]

A Model for Others?

The Partnership for Clean Fuels and Vehicles arising out of the 2002 World Summit on Sustainable Development sees the success in sub-Saharan Africa as the first step in the global phaseout of leaded gasoline. It launched a plan in 2006 to eliminate leaded gasoline from the rest of the developing world by 2008, including economies in transition, with an initial focus on the countries of the Middle East, North Africa, and western Asia. The partnership will then face one of its greatest challenges: the small island states of the Pacific.[9] Assuming the partnership's campaign succeeds, the world will be finally rid of the ecological shadows of Midgley's tetraethyl lead—85 years after the first public sale of ethyl gasoline.

10

The Globalization of Risk

Our understanding of the health effects of tetraethyl lead has come a long way since Thomas Midgley rubbed it into his hands in 1924 to demonstrate its safety. Repeated exposure to trace amounts of lead, the medical community is now sure, can cause lasting harm. It can reduce fertility and increase sperm abnormalities as well as contribute to premature births and low birth weights. It can impair children's brains and nervous systems as well as increase the risk of heart attacks and strokes in adults. And it can impair impulse controls and social skills, even contributing to delinquency in children and criminal behavior in adults.[1]

Scientists as far back as the 1920s thought adding a known poison to gasoline was risky; in light of such scientific uncertainty, some like Yale's Yandell Henderson called for precaution to prevail over corporate profits. In Henderson's opinion, industry-funded experiments "proving" that exposure to leaded tailpipe exhaust was "safe" were disingenuous right from the start, revealing nothing about the long-term consequences for people's health. For Henderson, it was only common sense that putting lead into gasoline could, and probably would, poison people "insidiously" as more and more cars jammed city streets—and that it would take many decades for the consequences to appear. By then, with millions of cars running on leaded gasoline, corporate and consumer resistance would mean long delays before the U.S. government could get rid of it.

The automotive industry saw such calls for precaution and further research as misguided, even hysterical. In early 1926, the U.S. Surgeon General gave the go-ahead to DuPont, General Motors, Standard Oil, and the Ethyl Gasoline Corporation. Sales took off; by the 1930s, some 90 percent of U.S. drivers were using this "superior" gas. By then, with industry scientists like Robert Kehoe controlling the research on leaded gasoline, U.S. policy makers had long forgotten Henderson's warnings. Several decades would pass before the geochemist Clair Patterson would again call into question the safety of leaded gasoline. His groundbreaking 1965 analysis of lead in humans in the northern hemisphere ignited a firestorm of controversy within the scientific community. Soon, study after study was confirming what Henderson had foreseen: leaded gasoline was indeed an insidious poison.

And, just as Henderson had also foreseen, it would take decades for the U.S. government to get rid of leaded gas. With so much profit to be made from this poison, corporations strove to discredit critics and delay regulations. And when at last there was a phasedown of leaded gasoline in the United States, it came with a consequence Henderson had not foreseen. To recover lost revenues as sales fell at home, U.S. corporations began to export more of their lead additive overseas, using the profits to diversify operations at home. Moreover, the economy of automobiles was now more international—with firms from many other countries also profiting from the sale of leaded gasoline—making any phaseout far more difficult than even Henderson could have predicted.

The result was an unequal—and duplicitous—global phasedown, starting in the early 1970s, with some states pursuing phaseouts of leaded gasoline while their multinationals were busy increasing sales elsewhere. Wealthy states, such as Japan and the United States, were generally able to take action sooner, although there were notable exceptions: poorer countries like Brazil and Thailand, for example, phased out leaded gasoline before richer ones like Britain and Australia. For governments that delayed taking action, phaseouts were often faster and cheaper—with governments in some developing countries receiving international aid to that end. Still, the basic trend was consistent: strong economies got rid of leaded gas first. Thus, by the start of the twenty-first century, 80 percent of the states still using this gasoline were in the developing world.

Since that time, the international community's record has improved. Indeed, international donors and cooperative corporations made the rapid phaseout of leaded gasoline in sub-Saharan Africa from 2002 to 2006 possible. A wave of phaseouts is now sweeping across the rest of the developing world, and, before long, only trace amounts of Thomas Midgley's leaded legacy will remain.

What explains this global pattern of change? Why did some governments react quickly and others slowly?

The globalization of environmentalism—in particular, worldwide research on the health effects of lead—explains some of this change. A global consensus about the dangers of leaded gasoline was solidifying by the 1980s. More and more governments began to see lead in human blood as unnatural and leaded gasoline as the primary source. More and more governments began to see the long-term health costs of leaded gasoline as far greater than the short-term economic benefits. Switching to unleaded gasoline thus became the sensible thing to do.[2] This growing global consensus empowered environmental agencies such as the U.S. Environmental Protection Agency at a time when the capacity of the lead industries to retain political and corporate allies was weakening (as reflected in their failure to derail the EPA's phasedown of leaded gasoline during the Reagan administration).

No doubt, then, regulations to lower lead levels in gasoline accelerated phasedowns in many countries. Yet, in countries with expanding markets for cars with catalytic converters, it's easy to exaggerate the influence of global environmentalism and national regulations. In the United States, where standards to lower the amount of lead in the gasoline pool were relatively strict, the turnover in the national auto fleet to vehicles with catalytic converters accounted for much of the decrease in leaded gas sales. One economic analysis of EPA policies from 1979 to 1988 put the accelerating effect of stricter lead-per-gallon standards and lead trading and banking options for refiners within the United States at "a few years" over "what fleet turnover would have achieved" by itself.[3]

Fleet turnover was a decisive force of change in most other countries as well. Over 90 percent of new cars now sold worldwide have catalytic emission controls; to date, automakers have sold over a half-billion vehicles with such controls. The profits just from vehicle emission control devices are considerable. The world market for these devices was over $48 billion in 2005—a market the Manufacturers of Emission Controls Association predicts will expand to over $70 billion by 2010.[4]

At the same time, rising global sales of automobiles and unleaded gas ensured that falling global sales for leaded gas would not hurt the profits of the world's major oil or auto companies. But those same sales hit the Ethyl Corporation hard. No longer a global powerhouse after GM sold it in 1962 to Albemarle Paper, Ethyl was able to delay phaseouts in some countries but did not have the clout to prevent the global shift toward unleaded gasoline.

The history of leaded gasoline peels away some of the layers of complexities of how and why ecological shadows of consumption form, intensify, and fade. It shows how science can both cause and mitigate these shadows, how over many generations the courage of a Patterson in 1965 can balance the ingenuity of a Midgley in 1921. It shows how corporations can hold scientific research captive, how the pursuit of profits and markets can silence calls for precaution in cases of high uncertainty. It shows how global trade can cast ecological shadows across the planet even during phasedowns of unsafe consumer goods in exporting countries—and how these shadows deflect environmental costs onto poor people living in unstable societies and ecosystems. On the other hand, however, it also shows how the same globalizing forces can promote environmental reforms. Thus rising global sales of catalytic converters and unleaded gasoline can explain much of the progress toward eliminating the ecological shadows of leaded gasoline. Environmentalists were able to cooperate with oil companies, automakers, and state agencies even in the poorest regions of the world, bringing knowledge, funds, and technologies to accelerate phaseouts.

The Midgley Genius

This history shows, too, how a single discovery can flow through a global political economy to harm billions of people. Thomas Midgley Jr. was hailed as a genius during his lifetime. In different times, he could well have been one of the inventors of the catalytic converter. A Ph.D. in mechanical engineering from Cornell was not enough for such a creative mind, and, after teaching himself chemistry, he went on to become a prolific inventor with wide interests, eventually holding over 100 patents. Ideas bounced around in his mind like balls in a pinball game. Nine out of every ten were "screwy," as one of his associates put it—like his idea to develop a golf ball able to soar over a mile off the tee. But the tenth was often brilliant. Firms like General Motors and DuPont were able to channel his eccentricities, using what GM's Charles Kettering liked to call "trial and success" to turn Midgley's genius into consumer goods. Midgley died in 1944, when he was just 55—strangled by a harness contraption he'd invented to lift himself out of bed after polio had left him partially paralyzed four years before.[5]

Generations later, the world is still struggling to cope with the consequences of Midgley's genius. His first great discovery, in 1921—using tetraethyl lead as an antiknock additive in gasoline—he knew was risky even at the time. But his second, in 1928—finding a stable, odorless, nontoxic, and nonflammable refrigerant for refrigerators—everyone could agree made for a much safer world (the toxic refrigerants then in use had a tendency to explode). It would take more than 40 years before anyone would question the environmental safety of the seemingly innocuous chemical compounds called "chlorofluorocarbons" (CFCs)—the subject of part III.

III

Refrigerators

11

Refrigerating the Ozone Layer

It took Thomas Midgley just three days in 1928 to discover a stable chemical compound to cool refrigerators, later sold by DuPont and General Motors under the trademark "Freon." All tests showed this new compound to be utterly safe—so safe, as Midgley would show the American Chemical Society in 1930, a person could breathe it in and blow out a candle. Before long, manufacturers everywhere were using similar compounds in a class known as "chlorofluorocarbons" not only as refrigerants in refrigerators and air conditioners, but also as propellants in aerosol spray cans and fire extinguishers and as foaming agents in foam insulation.

By the time anyone would begin to worry about the environmental consequences of consuming CFCs, they would be depleting the ozone layer, a consequence so startling it was beyond the hypotheticals of any scientist back in the 1930s. The silence of science for nearly a half century was not the result of industry control over research, as it was for leaded gasoline. Nor did Freon cause a tempest of controversy—or even calls for caution—as did ethyl gasoline. The specialists felt certain that Freon was an advance for consumer safety, replacing substances prone to explode and poison people.

The 1974 theory put forth by Mario J. Molina and F. Sherwood Rowland of how and why CFCs *could* deplete ozone in the stratosphere was an inspired one—more than worthy of the 1995 Nobel Prize in Chemistry.[1] The ignorance that prevailed before 1974 shows how science can simply get it wrong, how it can concentrate on fragmentary parts of a problem and miss the complicated whole. Yet the scientific research on ozone depletion during the 1970s and 1980s also shows how science can discover causes of complex global ecological change. This period reveals as well how this way of knowing can contribute to a global consensus to eliminate ecological shadows in the global commons (rather than, as with leaded gasoline, the shadow effects within sovereign territories). In this case, even with firms like DuPont challenging and delaying research, a decade after 1974 governments were well on the way to negotiating a series of international agreements to phase out the CFCs depleting the earth's ozone layer.

Many books survey the impact of these agreements on the consumption of CFCs. The next chapters take a different tack, narrowing the focus to the consequences for refrigerators. Chapters 11 and 12 analyze why these agreements were able to phase down the production and consumption of CFC refrigerators, while chapter 13 explores the environmental impact of replacing CFC refrigerators with "superior" CFC-free models—including, for example, the consequences for global energy consumption, natural resource use, recycling, and waste.

This approach allows for a more rounded analysis of how international agreements and new technologies interact with corporations and trade to change the ecological impacts of global patterns of consumption. It shows, again, how international law can accelerate efforts to replace consumer products harming the global environment with safer substitutes. It shows how international financial assistance can enhance the capacity of governments, firms, and consumers in developing countries like China and India to meet international environmental commitments. And it shows how, following international agreements, competition among corporations for trade advantages and shifting markets can improve the environmental efficiencies of producing, using, and replacing consumer products, with positive spillovers beyond just meeting international commitments.

The resulting efficiency gains per unit consumed can be significant. Yet such changes tend to rely on—and kindle—rising consumption, a fact that helps explain why the ecological shadow of the global refrigerator industry remains intense. As the following chapters will reveal, having all but phased out ozone-depleting gases, the industry is drawing down more natural resources, generating more waste, and producing more greenhouse gases the deeper it moves into the developing world. The analysis of this case begins by looking back at the refrigerator industry in North America over the first half of the twentieth century—a necessary step for understanding the initial reaction of governments, firms, and consumers to calls in the 1970s to phase out CFCs.

From Icebox to Gas Box

Down through the ages, people have kept food from spoiling in many inventive ways, from salting, drying, and smoking to storing in cellars, streams, and ice. Iceboxes, which allowed for longer storage during hotter months, did not become common in North America until the 1800s. These wooden cabinets—with ice on the top or bottom—were typically insulated with cork, sawdust, or seaweed and lined with tin or zinc. They were imperfect devices, requiring regular supplies of ice and subject to leaks, slime, and mice.

The search for a more efficient and reliable cooling unit for food was in full gear by the early twentieth century. General Electric began to market a machine in 1911 able to cool air by compressing chemical gases; a decade later, some 200 different refrigerator models were on sale in the United States. General Motors entered this emerging market by purchasing the Guardian Frigerator Company in 1918. Although some thought this an odd decision for an auto firm, there was a consistent logic to it: as with automobiles, GM could see a vast untapped consumer demand. Still, it was a risky investment. Like those of other firms, the refrigerators made by the Guardian Frigerator Company were bulky, wooden contraptions, unreliable, and susceptible to poisonous and smelly leaks. And the company had sold a mere dozen or so in two years of production, partly because the price tag, more than $700 ($11,000 in today's money), was well beyond the means of most consumers. At the time, GM was clearly far from its goal of putting an affordable refrigerator in every kitchen.[2]

Frigidaire

General Motors renamed the company "Frigidaire" and put Billy Durant in charge. Although Durant managed to raise sales over the next few years, the new models were still unreliable and expensive. By 1920, a discouraged Durant was ready to give up, when GM vice president Charles Kettering persuaded him to stay on. (As we saw in chapter 7, Kettering had been instrumental in Midgley's discovery of leaded gasoline in 1921.) With Kettering's backing, Frigidaire began to reduce the weight and size of its refrigerators. The company also became the first to start marketing more appealing metal cabinets (coated in porcelain). By the mid-1920s, Frigidaire was advertising these as cheaper, safer, cleaner, and more convenient—a must for any "modern" kitchen.

Frigidaire began to prosper. So did other manufacturers of mechanical refrigerators like Westinghouse and Kelvinator. As these firms began to mass-produce units with increasing efficiency, prices began to fall—from an average of $600 in 1920 to $275 in 1930—and sales began to rise. In 1925, 75,000 new refrigerators were sold in the United States; just five years later, the number had grown to 850,000. During the second half of the 1920s, most major refrigerator firms in the United States were experiencing annual increases in sales from 25 to 75 percent.[3]

Still, Durant and Kettering were unhappy with their refrigerators' cooling systems, whose toxic, corrosive refrigerants prompted some city authorities to require labels warning consumers of the hazards. It was now obvious to both men that the next breakthrough in the mass-marketing of refrigerators would require safer cooling systems. In 1928, Kettering turned to Midgley, and, with the assistance of Albert Henne and Robert McNary, Midgley discovered CFC-12.[4]

Freon

In 1930, DuPont (holding 51 percent) and General Motors (holding 49 percent) formed the company Kinetic Chemicals to manufacture CFC-12 under the trademark "Freon."[5] Inert and odorless, Freon was heralded as a great advance in refrigeration. At first, Frigidaire's competitors tried to taint consumer reactions to Freon with reminders that it contained toxic fluorine. But this had little impact. Scientists were united: Freon was safe. And consumers could see it was far superior to earlier refrigerants. By the mid-1930s, all of the major refrigerator companies were purchasing Freon (or licensing the right to make it). Refrigerator prices continued to fall during this period, dropping on average to $165 in 1935 and to $155 by 1940. Sales took off at an even faster pace. By 1937, over 2 million Americans owned a refrigerator. Just 20 years later, 8 out of 10 households had one.[6]

Meanwhile, after World War II, sales of CFC refrigerators also expanded quickly in other (primarily developed) countries with rising personal incomes and reliable household electricity. Hundreds of millions of CFC refrigerators and CFC freezers were in use by the time scientists began to question the environmental safety of a chemical safe enough to blow out a candle.

A Theory in *Nature*

Gracing the cover of the June 1974 issue of the British journal *Nature* were drawings of Galapagos birds—a blue-footed booby, a Galapagos penguin, a flightless cormorant, and a magnificent frigate bird, among others—by artist Hilary Burn. Inside this seemingly typical issue was an atypical article: a stunningly original two-page piece having nothing specifically to do with either birds or the Galapagos Islands. The title was hardly memorable: "Stratospheric Sink for Chlorofluoromethanes: Chlorine Atom-Catalysed Destruction of Ozone." The authors, Mario Molina and Sherwood Rowland, were barely known outside the chemistry community. Rowland was a well-regarded 47-year-old professor of chemistry at the University of California at Irvine. Molina was a 31-year-old postdoctoral fellow, who had only joined Rowland's research team a year earlier.

The theory first advanced in their article would one day change global relations.[7] The core idea was easy enough to follow. The use of chlorofluorocarbons had been expanding exponentially over the previous two decades.[8] Now, with millions of metric tons in use, and with producers adding more every year, an increasing amount was leaking from refrigerators, air conditioners, foam insulation, spray cans, and fire extinguishers. This might not seem worrisome, for CFCs were chemically inert in the earth's lower atmosphere, or troposphere, where weather occurs. But, Molina and Rowland reasoned, the same properties making them so stable could mean that more and more would drift up into the stratosphere with each passing decade. Here, they reasoned further, the intense ultraviolet radiation of the upper atmosphere could, at least in theory, break the chemical bonds of CFCs. Split apart, free chlorine atoms would then trigger a chain reaction that would destroy ozone (by removing one of the three oxygen atoms in each of its molecules). Such a process, if it did occur, would steadily deplete the ozone layer 20–50 kilometers (12–30 miles) above the earth. Because this layer protects the lower atmosphere and the planet's surface against the harmful effects of ultraviolet radiation from the sun, its destruction would be catastrophic for life on earth.

Although Molina and Rowland had no evidence to support their claims, at a press conference during the 1974 meeting of the American Chemical Society, they made the natural leap from theory to policy. Because "the risks that are involved are too large," explained Rowland, "we ought to discontinue putting chlorofluorocarbons into the atmosphere." He pointed out that, if their theory were correct, 1 percent of the ozone layer was already gone. Swift action was vital. If CFC production were to continue to rise, 7–13 percent of the ozone would disappear in a century or so. The consequences for crops, the global climate, and human health would be devastating. Rates of skin cancer and cataracts would rise. Even losing 5 percent of the ozone layer could cause a 10 percent increase in skin cancer rates.[9]

Representatives from DuPont, the world's largest producer of CFCs, listened politely as reporters scribbled notes. Then, for the very first time, DuPont responded. Yes, CFCs were drifting around the lower atmosphere and sinking into the oceans (DuPont had funded some of the research showing this). But no one had ever detected a single CFC in the stratosphere. How could CFCs reach such heights? Molina and Rowland's theory was all conjecture—little more than the imaginative musings of ivory tower academics.

Where, DuPont demanded, was the *evidence*?[10]

Collecting and Contesting the Evidence

In July 1975, DuPont announced a multimillion-dollar research effort to test this theory. "We are trying to find the truth," wrote Irving S. Shapiro, DuPont's chairman of the board, in the *Washington Post*. "There are some who say that aerosols should be banned now even before the facts of the studies are known. DuPont wants to do what is right—for people, for the aerosol industry, and for ourselves—but we believe sincerely there is time to gather information and make a reasoned decision."[11]

Others were soon conducting tests as well. Balloons sent into the stratosphere by the U.S. National Oceanic and Atmospheric Administration (NOAA) in 1975 found CFCs above 19 kilometers (12 miles), in "close agreement" with scientific predictions.[12] Firms like DuPont were now, in the language of Rowland, "mobilizing"—deploying tactics to create uncertainties and delays.[13] The corporate mantra was in place by the autumn of 1975: "Before a valuable industry is hypothesized out of existence, more facts are needed."[14]

After a 1976 study by the National Academy of Sciences found clear evidence supporting Molina and Rowland's theory, the United States did indeed ban the use of CFCs in aerosol spray cans in 1978, in part because affordable substitutes were readily on hand. A few other developed countries, as chapter 12 will elaborate, also took unilateral action to reduce CFC use in aerosol spray cans. But little was done to prevent the rising consumption of CFCs in other products, particularly refrigerators and freezers.

The governing council of the United Nations Environment Programme first raised the issue of ozone depletion at the international level in 1976.

The following year, UNEP joined forces with the World Meteorological Organization (WMO) to begin assessing depletion rates. Intergovernmental negotiations for an international agreement on phasing out ozone-depleting substances started in 1981. Progress was slow in the first half of the 1980s, by which time over 100 million CFC refrigerators were operating in the United States alone. Total annual CFC production was rising, too, with the United States accounting for 30 percent, when, in 1985, British scientists found a giant "hole" in the ozone layer over the Antarctica—as big as North America and lasting three months. Finding this hole was a turning point and spurred an emerging consensus on the need for quick action.[15]

States Act Globally

The Vienna Convention for the Protection of the Ozone Layer, a framework convention without legally binding targets, was adopted in 1985. Two years later, the Montreal Protocol on Substances That Deplete the Ozone Layer was adopted to set binding targets to reduce the production of ozone-depleting CFCs and halons.[16] In 1988, a year before the Protocol went into force, DuPont, responsible for between 20 and 25 percent of global CFC production at the time, announced it would move ahead with marketing affordable substitutes.[17] Then, in 1990, the governments of developing countries that were parties to the protocol signed the London Amendments to phase out consumption of CFCs and halons by 2010 (with "consumption" defined as production plus imports minus exports). The London Amendments set the year 2000 as a phaseout deadline for eight CFCs in developed countries—a date moved forward to 31 December 1995 in Copenhagen in 1992. Conferences in Montreal in 1997 and Beijing in 1999 further amended and strengthened the Montreal Protocol, adding other ozone-depleting substances and accelerating the phaseout schedules.

Results came quickly. By the mid-1990s, developed countries were no longer producing CFCs; by the second half of the 1990s, the developing world was steadily reducing its production as well. Globally, CFC production fell from a peak of nearly 1.1 million metric tons in 1987 and 1988 to just 80,000 metric tons by 1996. Over the next decade, progress was also made in reducing CFC use in the developing world, and, by 2003, the world was producing less than 20,000 metric tons of CFCs.[18]

Many factors made this international effort "a striking success."[19] The scientific community was able to collect compelling evidence within a decade after Molina and Rowland first published their theory. The causes (CFCs and halons) and consequences (skin cancer and cataracts) of ozone depletion were reasonably clear-cut, as was the solution (replace CFCs)—at least compared to the solution to climate change or deforestation. The small number of firms responsible for CFC production at the time of the Montreal Protocol—just 21 in 16 countries—made global negotiations considerably more manageable. Developed countries accounted for 88 percent of production, with only a handful of chemical producers in leading roles, notably, DuPont, Imperial Chemical Industries (ICI) of the United Kingdom, and Atcham (a subsidiary of Elf-Aquitaine) of France.

Explaining the "Success"

Thus the willingness of corporations and states profiting from CFCs to accept the theory and evidence of ozone depletion explains much of why the international community was able to negotiate—and then implement—binding targets to reduce CFC production. But why were these firms and states so willing? One reason was their strengthening environmental commitment. International financing, which helped bring some states and firms in the developing world on board, was another. But the primary reason was the capacity of the world order to substitute CFC-free consumer goods and produce them in ever greater numbers: global trade kept expanding, foreign investors kept competing, corporate profits kept rising, economies kept growing, and choices for consumers kept improving. The emerging political economy of CFC-free goods was able to counter resistance from the old political economy of CFC goods, as chapter 12, on the global phaseout of CFC refrigerators, will show.

12

Phasing Out CFC Refrigerators

The global solution to CFC refrigerators has been simple: replace them with even more CFC-free refrigerators. International agreements and aid have spurred along this process of environmentally friendly change, with state regulators and corporate codes of conduct helping to ensure compliance and consistency. But, as this chapter will show, the main force of change has been the global market for CFC-free refrigerators, which has afforded firms opportunities to invest, trade, and profit from growing sales of these refrigerators—and thus recoup any revenue lost from phasing out CFC refrigerators.

The phaseout of CFC refrigerators in the First World started shortly after the Montreal Protocol went into force in 1989, accelerating quickly in the first half of the 1990s after chemical companies like DuPont began marketing effective and affordable CFC substitutes and appliance firms began to concentrate on capturing markets for CFC-free refrigerators. Although this process involved some exporting of environmental costs to the developing world (such as dealers shipping used CFC refrigerators to Africa), this was nothing compared to the wholesale exporting of costs in the case of leaded gasoline. Why the difference?

The international commitment to eliminate CFCs explains some of the difference. And the scope of the consequences explains some of it as well: lead poisoning from leaded gasoline affected primarily local populations in developing countries, whereas rising skin cancer rates from depletion of the ozone layer affected people worldwide. But, as this chapter argues, much of the difference is simple economics: in the case of leaded gasoline, it made financial sense to expand sales in developing countries, whereas, in the case of CFC refrigerators, it did not. Instead, it was more logical for appliance companies to trade with—and invest in—these countries to capture emerging CFC-free markets. It was more logical, too, for local producers in developing countries to pursue joint ventures, export markets, and international environmental financing for CFC-free refrigerators, as exemplified by the expansion of China's refrigerator industry since the early 1990s.

Worldwide, by 2000, CFC-free refrigerants had replaced CFCs in 95 percent of the world's refrigerators. Almost all of this change occurred during the 1990s. Not having cost-effective substitutes readily at hand, companies like DuPont and Whirlpool strongly resisted change in the 1970s and 1980s. This explains in large part why, without an international agreement, the initial response to the emerging scientific consensus on the risks of CFCs failed to make much headway—rising sales of items that didn't have low-cost substitutes for CFCs (refrigerators, air conditioners, foams, and cleaning solutions for electronics) all but offset sales of items that did (hairsprays and deodorants).

CFCs in the 1970s and 1980s

Aerosol propellants accounted for about 60 percent of the worldwide use of the two primary chlorofluorocarbons—CFC-11 and CFC-12—in the mid-1970s. Various governments took unilateral steps in the 1970s and early 1980s to phase these out. Some, like Norway, the United States and Sweden, banned the use of CFCs as propellants except for essential uses. Others, like Canada, imposed bans on aerosol sprays for cosmetic, drug, and hygienic products. Still others, such as countries within the European Community, called for a voluntary reduction of 30 percent in the use of CFC-11 and CFC-12 in aerosols.

Governments took such steps to slow the growing production of CFCs—and thus gain time to gather data on the consequences for the ozone layer. These efforts were moderately successful. The amount of CFCs in aerosols fell steadily in the second half of the 1970s and early 1980s. The total amount of CFC production and sales went down over this time as well, although only slightly as sales of CFCs for other uses climbed. The total began to rise again during the 1980s, however, with the decline in aerosols leveling off and with the use in other sectors continuing to rise. By 1987, the total amount of global sales of CFC-11 and CFC-12 was not much different than in 1974. But, by then, production of the third most significant chlorofluorocarbon (CFC-113) was far higher, exceeding 200,000 metric tons per year for the first time.[1] Still, without the efforts to reduce the use of CFCs in aerosols, the situation by the time of the Vienna Convention and Montreal Protocol would have been worse. Although aerosol sprays still accounted for 300,000 metric tons of CFCs in 1986, that amount represented only a little more than one-quarter of "controlled use" worldwide.[2]

In contrast to reducing CFCs in aerosols, little effort was made between the mid-1970s and mid-1980s to reduce the use of CFCs in refrigerators, freezers, or air conditioners. Refrigeration and air-conditioning accounted for between 25 and 30 percent of global CFC use by the mid-1980s.[3] Here opposition from industry was a core reason for the failure to regulate CFCs. This resistance began to wane by the mid-1980s, when many U.S. firms, including DuPont, changed tack, supporting the Montreal Protocol and sincerely pursuing CFC substitutes.

Phasing Out CFC Refrigerators in Developed Countries

The Montreal Protocol was a turning point for the refrigeration industry. The world's top three manufacturers of refrigerators and freezers—the Swedish firm Electrolux, the U.S. firm Whirlpool, and the German firm Bosch und Siemens Hausgeräte (BSH)—all took steps in the late 1980s and first half of the 1990s to retool plants to handle CFC substitutes, recycle or retrofit old refrigerators, and train workers to avoid venting CFCs during servicing. Electrolux replaced ozone-depleting and climate-changing substances in the early to mid-1990s in Europe and North America. BSH halved its use of CFCs in 1988, and ended its use of them and of hydrofluorocarbon (HFC) greenhouse gases in its European plants in 1993 (replacing these with the hydrocarbons isobutane and cyclopentane).[4] The refrigeration industry was not exceptional here. By the start of 1996, no developed country was producing the main CFCs (11, 12, and 113), except for a small amount set aside for essential uses. World production of the five main CFCs (11, 12, 113, 114, and 115) was less than 80,000 metric tons in 1996. Although a small quantity of CFCs was smuggled into these countries (or, sometimes, illegally traded within), in the end, jurisdictions like the United States and the European Union "overcomplied" with the rules of the Montreal Protocol.[5]

The successful phaseout of CFC refrigerators involved some transfers of harm into the developing world. Some European dealers, for example, were shipping used CFC refrigerators to Africa in the second half of the 1990s. Some consumers refused to buy these imports, even though they were much cheaper than CFC-free refrigerators (costing half as much in Zambia, for example). But others, as you'd expect anywhere, saw a bargain and didn't seem to worry much about the consequences for the global environment. "Whatever this CFC means," explained one

Zambian fishmonger in 1999, "all I know is that the deep freezer has been a big asset for me." "This is just one small fridge," rationalized a Zambian civil servant in the process of buying a CFC refrigerator in 1999, "surely it can't be compared to those big factories which smoke heavily. I doubt if it can really have much impact at all."[6]

The dumping of used CFC refrigerators in the 1990s certainly made it harder for some developing countries to meet their international commitments. Poor maintenance, servicing, and recycling of refrigerators in these countries also meant CFCs were commonly going directly into the atmosphere. At the same time, however, the changes in CFC use within the developed world were altering global markets, technological capacity, and multinational incentives in ways conducive to phasing out CFC refrigerators in the developing world. A consensus, moreover, was emerging among First World states over the need to provide the Third World with financial support for these phaseouts.

The Multilateral Fund

The phaseout of CFC refrigerators took on a more global character after developing countries in the Montreal Protocol agreed in 1990 to a target of 2010 for a phaseout of the two most common ozone-depleting substances: CFCs and halons. The following year, the Multilateral Fund for the Implementation of the Montreal Protocol was established to assist developing party states with meeting their commitments. Financed by developed states and economies in transition, the fund had an initial budget from 1991 to 1993 of $240 million. Since then, it has been replenished five more times with sums of between $400 and $475 million; its 2006–2008 budget was $470 million.

Only countries whose annual consumption and production of ozone-depleting substances was below 0.3 kilograms per capita (30 metric tons per 100,000 people) qualify for assistance. There were 145 eligible countries (of the 191 Parties to the Montreal Protocol) as of November 2007. By then, the fund had approved over $2 billion through four implementing agencies: the United Nations Environment Programme (UNEP), the United Nations Development Programme (UNDP), the United Nations Industrial Development Organization (UNIDO), and the World Bank. The fund estimated that its financial support has been essential for phasing out 190,000 metric tons of consumption and 116,000 metric tons of production of ozone-depleting substances by the beginning of 2006.

Thus financing from this fund has been instrumental in reducing CFC use in the refrigeration sector in countries like China and India. The Indian government ratified the Montreal Protocol in 1992. To meet its protocol commitments by 2010, it's working to phase out refrigerators using CFC-12 gas. According to the Indian Department of the Environment, fewer than 30,000 such refrigerators were still in use by 2005. With financing from the Multilateral Fund, the Indian government has been purchasing kits to convert these to cooling systems using liquid petroleum gas (LPG), which does not release any of the chlorine—or bromine—responsible for ozone depletion.[7]

The Case of China

China ratified the Montreal Protocol in 1991, a year before India, when CFC-11, CFC-12, and halon-1211, used in aerosols, foams, air conditioners, solvents, fire extinguishers, freezers, and refrigerators, accounted for over 90 percent of China's total production and consumption of ozone-depleting substances. By the mid-1990s, as a result of declines across the developed world, China was the world's largest producer and consumer of these substances, accounting for about one-third of the global total.

By then, firms within China were producing some 12 million refrigerators and freezers a year, accounting for around half of all of the refrigerators made in developing countries belonging to the Montreal Protocol. Of these, about 3 million units were being built with technologies using less ozone-depleting substances (some units were even CFC free). The manufacturing lines of another 5 million refrigerators were already in the process of retooling to reduce the reliance on ozone-depleting substances. In China, as in India, the Multilateral Fund has supported efforts to shift to CFC-free cooling systems over the last decade and a half, having disbursed over $75 million to 49 projects in the household refrigeration sector by the middle of 1998. A total of some 2,500 metric tons of ozone-depleting substances had been phased out using these funds by the beginning of 1998.[8]

Yet, as was true elsewhere in the developing world, international financing by the Multilateral Fund was just one of many factors motivating the decrease in CFC use in China. China's decision to ratify and later comply with the Montreal Protocol to gain international legitimacy was a factor, as was maneuvering by China's main implementing agency to expand its influence.[9] But market demand, including access to overseas markets and technologies, has been a particularly strong factor—with some, like environmental researchers Jimin Zhao and Leonard Ortolano, finding it to be the "most significant" motivating factor for household refrigerator and freezer producers within China. "If we could obtain a good market share [for CFC-free refrigerators]," one manager explained, "we could conduct ODS [ozone-depleting substances] reduction even if there were no financial support [from the Multilateral Fund]. But if we could not obtain a good market share, we would not carry out ODS reduction even if financial support were available."[10]

With old markets closing and new ones in sight, a few local refrigerator firms in China were voluntarily adopting more expensive CFC-free cooling systems even before China ratified the Montreal Protocol in 1991. China's total refrigerator exports fell 58 percent from 1988 to 1991, partly because, after so many European governments signed the protocol in 1987, more and more consumers in Europe were turning to CFC-free refrigerators. By 1991, China was managing to export only 230,000 refrigerators a year—just 4 percent of its total production. Because the protocol only allowed trade among parties, ratification by China would open up new overseas markets for CFC-free refrigerators (which explains why many firms in China urged the government to ratify).

Efforts by China's 40 refrigerator manufacturers to replace ozone-depleting substances gained momentum after 1991. Through the 1990s, some began using technologies requiring smaller amounts of CFCs (generally about half as much). Others began to use transitional substitutes—such as hydrochlorofluorocarbons (HCFCs)—with smaller ozone-depleting impacts than CFCs (although with significant global warming effects). And a few manufacturers switched to CFC-free systems. By the beginning of 1998, projects funded fully by industry had phased out just 200 metric tons less ozone-depleting substances than the reductions arising from Multilateral Fund projects.

Hoping at first this would get them back into the European market, some Chinese refrigerator firms urged their government to adopt environmental labeling to assure consumers that a refrigerator was either CFC free or 50 percent CFC reduced. These labels first appeared in 1993. Before long, however, refrigerator firms within China were employing environmental labels to gain a marketing edge over competitors. By 1996, many of the leading firms were advertising the value of "world-class" CFC-free refrigerators. Some were even misleading consumers, implying a CFC refrigerator in a home could directly harm a person's health. The early results of the marketing campaign were promising: more than enough consumers were willing to pay from 10 to 15 percent ($25 to $45) more for a CFC-free refrigerator. Just two years later, 27 of China's 40 refrigerator manufacturers were using these labels.[11]

Multinational appliance firms also contributed to some of the decreases of CFCs in China. Appliance firms from countries like Sweden, the United States, Germany, Japan, and Italy also began to invest more in the Chinese production of CFC-free refrigerators after 1995. Some domestic manufacturers imported technologies to manufacture CFC-free refrigerators from overseas firms; others entered into joint ventures with them. Indeed, by the late 1990s, nearly a third of China's refrigerator manufacturers had entered joint ventures to produce CFC-free refrigerators.

In recent years, many multinational companies have publicized these joint ventures as evidence of their corporate social responsibility. Thus the German appliance maker BSH, in its 2004 annual review of its sustainability practices, made much of its initiative to manufacture only CFC- and HFC-free refrigerators at its joint-venture Chuzhou plant (opened in 1996) from 1999 on.[12] These joint ventures have helped to spread information and technologies for reducing CFCs among domestic Chinese refrigerator makers. They have added as well to market pressures to switch to CFC-free technologies. Some Chinese firms—in particular, state- and collectively owned ones—had neither the technical nor the financial capacity to compete with these more modern refrigerator plants and went out of business within a few years.

Over the 1990s, China made significant per unit progress in reducing ozone-depleting substances in refrigerators and freezers. By 1997, its total consumption of these substances had risen only moderately (some 1100 metric tons) even though refrigerator production had more than doubled (from 5.5 to 13 million units).[13] Although financing by the Multilateral Fund has accelerated the rate of environmentally friendly change, as this section shows, market demand and, to a lesser extent, multinational investments have been more significant. Such change has been harder to achieve in the foam and solvent sectors, which, unlike the refrigeration sector, do not have the advantage of a small number of firms with the capacity to import new technologies and to meet the basic criteria for financing by the Multilateral Fund.[14]

Still, across all sectors, the statistical story for the Multilateral Fund in China is one of progress over the last decade and a half. By mid-2003, the Multilateral Fund had given China $470 million for 403 projects, supporting a phaseout of 53,900 metric tons of production and 87,600 metric tons of consumption of ozone-depleting substances across all sectors. This represents about half of all decreases in ozone-depleting substances during the previous dozen years. The fund has now approved over $700 million to assist China, which is now "on track" to meet its commitment for a full phaseout of CFCs and halons by 2010.[15]

A Global Phaseout

Keeping major Third World producers and consumers like India and China on track to meet the 2010 targets—though critical to protecting the ozone layer—is only part of the solution. The smaller countries in the developing world must also phase out ozone-depleting substances. To help them do so, the Multilateral Fund is distributing its financing widely. So far, it has approved funding for over $2 billion for some 5,500 projects in 144 countries, involving activities such as industrial conversion, capacity building, technical assistance, and training. Having raised "the phase-out of CFCs in the refrigeration servicing sector in smaller countries" as an area needing special support at its April 2006 meeting, the fund's executive committee allocated $63 million to 47 developing countries to that end and urged implementing agencies to accelerate progress in this sector.[16]

Overall, then, the combined impact of new policies, targeted financing, and corporate positioning—first in developed countries, then in developing countries—has significantly decreased the volume of CFCs flowing from refrigerators and freezers over the last two decades. The progress in the 1990s, beginning in the wealthy world and then moving to all corners of the globe, was especially impressive, with firms and governments replacing CFCs in 95 percent of the world's refrigerating systems.[17] But this process has had its downside. Replacing CFCs with HCFCs and HFCs was, in the words of Greenpeace, "switching from disastrous to very bad" because both substitutes significantly contribute to climate change.[18] But, here as well, advances are now occurring worldwide in substituting more benign refrigerants (such as pure hydrocarbons).[19] New refrigerators and freezers, especially solar-powered ones, using these more benign chemicals are certainly better for the environment. Yet, although these appliances, like their CFC predecessors, help save lives and safeguard health by preserving necessities such as food and vaccines, because the total number of them continues to rise, their ecological shadow continues to extend across the planet.

Selling More Refrigerators

Phasing out CFCs under the Montreal Protocol was good for refrigerator sales. Advertising campaigns stressed the environmental value of upgrading to new, CFC-free models, which manufacturers began to market with many other enticing features (including, as chapter 13 documents, much higher energy efficiency). Some governments even compensated consumers for replacing old refrigerators with new ones.[20] Such changes and programs contributed to climbing worldwide sales of electrical appliances over the last decade. The number of units being sold annually went from 1.01 billion in 1998 to 1.26 billion in 2003. Of these, large kitchen appliances—such as refrigerators, freezers, stoves, washers, dryers, and microwaves—accounted for 252 million in 1998 and 319 million in 2003. Future sales are expected to be even higher. Indeed, sales of electrical appliances are projected to reach some 1.6 billion units in 2008—with large kitchen appliances accounting for more than 400 million of these.

Many other factors are stimulating appliance sales as well. Trading agreements such as the North American Free Trade Agreement (NAFTA) and the Southern Common Market (Mercosur) have helped to open markets for kitchen appliances. Barriers to trade remain in a few transitional economies and developing countries, but, even here, the trend is toward fewer taxes, duties, and restrictions on large foreign appliances. A program of liberalization in India, for example, now allows foreign firms to invest, market, and import large kitchen appliances. The Indian government removed quantitative limits on imports in 2000 and lowered custom duties in 2001 (from 40 to 35 percent). There are many other examples of this trend, such as Russia's decision in 2001 to lower customs duties on large kitchen appliances from 20–30 percent to 15–20 percent.

Higher per capita incomes, rising populations, and expanding electricity infrastructures have stimulated a steady growth in global sales for refrigerators and freezers: there were 82 million sold in 2002—up 9 million from 1998. Over two-thirds of these sales in 2002 were in the Asia-Pacific (28.8 percent), western Europe (23.7 percent), and North America (18.4 percent), with Latin America (10.9 percent), Africa, the Middle East (10.2 percent), and eastern Europe (7.1 percent) accounting for the rest.[21] Worldwide growth has been slowed somewhat by inadequate and unreliable power grids, which still limit consumption in emerging markets like China, India, Southeast Asia, and Latin America.

Competing to Expand Consumption

Markets for refrigerators are now, in the language of financial analysts, relatively "mature." Competition is stiff, mergers and acquisitions are common, and many small manufacturers and retailers are being forced out of business by increasingly large and complex multinational enterprises, which are relying more and more on the power of brand names and lower prices to fight for international market shares.

A few firms from developing countries, such as the Chinese refrigerator manufacturer Haier, are expanding abroad, moving into established markets like Japan and the United States. The competition is fierce here, especially from established brands like Electrolux, Whirlpool, and Hitachi. Still, Haier has managed to enter the U.S. market with a fast-selling line of small and cheap refrigerators through the Wal-Mart chain.[22]

In the aftermath of the government mandating CFC-free manufacturing of refrigerators by 2005, other Chinese refrigerator firms have joined Haier in seeking entry into global markets. Of the 30 million household refrigerators made in China in 2004, some 12 million, worth $777 million, were exported, primarily to the European Union, the United States, Asia, and the Middle East. This represented a 40 percent increase over 2003 exports, which were nearly 36 percent higher than exports the year before. Such rapid growth is creating pressures on Chinese refrigerator producers to merge or go under. Already, the number of locally owned refrigerator and freezer firms has fallen from about 100 to less than 40, and the sector is expected to consolidate at a half dozen or so firms.

Unlike the 1980s and early 1990s, when the main restructuring in China's refrigeration sector involved multinational corporations taking over or forming joint ventures with local firms, since then, the trend toward larger and fewer refrigerator firms has involved local firms acquiring or investing in other local firms. By 2004, China already accounted for about 30 percent of global output of household refrigerators; its output is expected to grow much larger as Chinese refrigerator firms work hard to expand sales at home and abroad.[23]

Major players like Electrolux, BSH, and Whirlpool are not sitting idly by. A key focus of these multinationals is on expanding markets in developing countries. Thus, in 2002, Whirlpool bought out its joint venture partner in China—Whirlpool Narcissus Shanghai—to expand its cheap manufacturing base for exports to the Asia-Pacific region. That same year, it established Whirlpool Mexico, providing direct access to the Mexican market as well as further export opportunities into the Caribbean and Central and South America, and, to expand sales in Central Europe, it secured a cheap manufacturing base there by acquiring Polar S.A. from ElcoBrandt, the leading appliance brand in Poland.[24]

Building new markets in the developing world is just one of many corporate strategies to expand consumption. These firms are also encouraging consumers to upgrade to new models (often larger or with more features) or to purchase a second or third appliance for the convenience of extra storage or more options.[25] Many factors influence such purchases. In some countries, demand for bigger refrigerators and freezers is rising as consumers switch from daily shopping to bulk and weekly shopping. Habits are changing for many reasons. More stores are offering deals in bulk, "Buy one, get one free," being just one of countless marketing schemes. Demand for more efficient ways to manage food shopping is increasing with longer working hours and more women working outside the home. The trend toward bulk buying is especially strong in countries like the United States, where homes are large and where microwavable and frozen foods are popular.[26] It's less so in countries like France, Italy, and Spain, where cultural traditions of meals made at home with fresh ingredients are stronger, or in countries like Japan or Taiwan, where homes are smaller and storage space limited.

The United States was by far the single largest national market for refrigerators and freezers for the period 1998–2002. Sales rose from 11.6 million units in 1998 to 14.1 million in 2002—a "healthy growth" of nearly 22 percent in volume. Growth of all "big-ticket" consumer items was helped along by low interest rates, long-term credit agreements, and rising personal disposable income (which grew nearly 24 percent, from $6.3 trillion in 1998 to $7.8 trillion in 2002). The world's second largest national market for refrigerators and freezers was China, with sales of 8.1 million units in 2002—down from a peak of 9.6 million in 1999 (insufficient electric power is hampering sales in rural areas).[27]

Governments applaud these burgeoning refrigerator sales as a sign of prosperity and a source of overall economic growth. But many now also recognize the pressures such rapidly rising consumption puts on ecosystems, as do increasing numbers of corporations. The global effort to replace CFC refrigerators is just one of many initiatives to decrease the environmental impact of the production, use, and disposal of refrigerators. Chapter 13 will analyze some of the other initiatives—with a focus on corporate efforts to market "superior" refrigerators.

13

Selling the "Superior" Refrigerator

Over the last two decades, competition among refrigerator and freezer companies has extended beyond just marketing CFC-free products. Firms like Electrolux, Whirlpool, and Bosch und Siemens Hausgeräte (BSH) have also been competing to develop more energy-efficient models, advertising them as win-win purchases for consumers—a way to save the environment *and* save on utility bills. Many are upgrading factories, modifying packaging, and introducing codes of conduct to conserve energy, emit less pollution, reduce waste, and promote recycling. Some are cooperating with governments and nongovernmental organizations to develop environmental legislation to make producers more responsible for recycling, thus creating incentives to develop models that are easier and cheaper to recycle (as well as justifying higher consumer prices across the refrigeration sector). A few firms, such as Electrolux, are even auditing suppliers in developing countries like China and Brazil to monitor compliance with corporate codes of conduct and environmental regulations.

Today, as a result of these environmentally friendly changes, a new refrigerator tends to use less energy and produce less waste than even a decade-old model. Every corporate brochure on social and environmental responsibility touts these commendable changes as "progress" for consumers and the global environment. But do they constitute *true* progress? The answer, this chapter argues, is a qualified no. Although the changes do save some consumers money on utility bills, they don't always lower personal consumption: consumers in "mature" markets tend to buy more refrigerators and freezers with more storage capacity, thus drawing more electricity and producing more waste. Nor have they translated into smaller impacts on the global environment: rising worldwide consumption of refrigerators and freezers offsets the per unit gains in producing, using, and disposing of them. Before considering these more critical points, however, let's survey corporate efforts to sell the new and "superior" refrigerators.

Marketing Energy-Efficient Refrigerators

Guidelines to promote the energy efficiency of home appliances are now common in many countries. Governments use many different tactics and policies to promote higher efficiency. The United States imposes minimum energy efficiency standards for appliances (with, for example, rules to require manufacturers to improve current efficiency over past efficiency in a baseline year). Belgium gives consumers cash bonuses, funded by its energy firms, for buying refrigerators with the highest energy efficiency rating. France relies on its state-owned energy firm to educate consumers about the value of buying an energy-efficient appliance. Some countries in the developing world—China, for example—have offered financial incentives to domestic producers to develop models with less environmental impact.[1]

In many households, running a refrigerator is the single largest expense on the electric bill. In the United States, for example, refrigerators and freezers account for about 14 percent of the electricity consumption of U.S. households: a typical refrigerator costs over $1000 to run over its lifetime (with operating costs rising as the refrigerator ages).[2] Many jurisdictions, including the United States, China, and the European Union, impose caps on the allowable level of energy consumption for refrigeration appliances. Many also require firms to include energy ratings to inform consumers. The European Union, China, and Australia, for example, all have mandatory energy labeling schemes for refrigerators, freezers, and air conditioners.

In 1995, the European Union adopted a mandatory rating system for refrigeration appliances. Ratings are from A to G, with A standing for most and G for least energy-efficient. The EU added two additional ratings for refrigerators and freezers in 2004: A+ and A++, earned for using 25 percent and 45 percent less energy, respectively, than an A-rated appliance. These new ratings are not easy to achieve. Only 18 percent of the refrigerators and freezers made by BSH, for example, were rated A+ or A++ in 2004; the following year, only 21 percent were.[3]

International programs also encourage manufacturers to develop appliances with higher environmental standards. The world's largest appliance firms all participate in these programs to some degree. The Swedish firm Electrolux, the world's leading manufacturer of refrigeration appliances by volume in 2002—selling under the brand names "Electrolux," "Frigidaire," "Zanussi," and "Kelvinator"—widely publicizes its policy of corporate social responsibility, as do the world's second and third largest refrigerator and freezer makers in 2002: the U.S. firm Whirlpool and the German firm BSH.[4]

Environmental Responsibility at Electrolux

Electrolux is one of Sweden's largest companies, with 276 subsidiaries in 60 countries and with sales of more than 40 million products in more than 150 countries. At the core of the environmental code of conduct for the Electrolux Group of companies, Electrolux's environmental policy commits its companies to improve the environmental performance of their suppliers, producers, users, and recyclers. This policy is implemented through "globally facilitated, locally owned management systems" that comply with the International Organization for Standardization (ISO) 14001 standard, the only certifiable standard in the ISO 14000 series.[5] Under the specific guidelines of the standard, a firm must (1) declare it will comply with all environmental regulations in its locale; (2) put a management system in place that conforms to its environmental policy; (3) commit both to preventing pollution and to continually improving environmental management; and (4) agree to encourage all contractors and suppliers to implement environmental management systems that meet the ISO 14001 standard.

All production units in the Electrolux Group with more than 50 employees must receive ISO 14001 certification. The number of units certified by ISO 14001 went from fewer than 20 percent in 1998 to nearly 80 percent by 2003. At the beginning of 2006, 68 Electrolux manufacturing units were certified—or 91 percent of the total number requiring certification and 98 percent of the Electrolux Group's total manufacturing capacity.[6] Electrolux is using this standard to create a baseline for monitoring compliance across its many units. Like the other 90,000 or so firms certified under this standard, it is also using ISO 14001 certification to reassure consumers of its environmental commitment (and thus sell more of its products).

As part of that environmental commitment, Electrolux is implementing policies to address climate change. Because electricity for home appliances accounts for about 4 percent of total carbon dioxide emissions in Europe, a primary tactic is to "stimulate" consumers to purchase more energy-efficient appliances. Electrolux Home Products, for example, cooperated with the Dutch energy company Eneco in a 2005 advertising campaign in the Netherlands on the value of buying more energy-efficient products. Electrolux is also working to improve the energy efficiency of its products, in recent years achieving an annual average efficiency gain of 4 percent across all products. "Too many homes have appliances running on technologies developed more than a decade ago," explains Environmental Affairs Vice President Henrik Sundström. "Although we continue to cut energy levels in our products, the best approach I know to cut consumption is to encourage customers to exchange models that are more than 10 years old for new, more efficient ones."[7]

Electrolux sees expanding sales of more energy-efficient appliances in China—a market that could represent as much as 35 percent of global demand for appliances by 2012—as critical for the success of this strategy. Its efforts to that end are aided by China's mandatory energy labeling scheme for refrigerators, freezers, and air conditioners. Electrolux is working as well to reduce waste from packaging. The World Packaging Organization, an international federation of packaging institutes, awarded the Electrolux freezer plant in Hungary a WorldStar award in 2004 for developing recyclable pressed-cardboard packaging that saves around 900 metric tons of wood a year. To ensure that its manufacturers comply with the Electrolux code of conduct (including its environmental policy), the company provides regular training programs for managers and employees on the code (as well as on business ethics more generally). It's also conducting audits: 12 in 2004, with 2 follow-ups and 1 new audit in 2005; by 2006, all Electrolux factories in Asia, and all but one in Latin America, had gone through an audit for social and environmental practices.

Because increasingly its products and components are made in developing countries, suppliers are a growing part of the global impact of Electrolux. Monitoring and enforcing its code of conduct among these suppliers can encounter thorny cultural and political problems, such as in China, one of the more difficult places to hold suppliers to foreign standards. Even here, however, Electrolux completed the first phase of its Supplier Monitoring and Compliance Program in 2005. This involved 45 audits of large and small firms to gain experience before finalizing the procedures for evaluating its 400 or so Chinese suppliers. The initial audits found many cases of "noncompliance," although Electrolux expects higher compliance once suppliers gain a better understanding of expectations. Over the next few years, Electrolux intends to assist suppliers in complying with the corporate code of conduct. "Ending a contract with a supplier does not improve the situation for workers or the environment," explains Jean-Michel Paulange, head of Electrolux purchasing for the Asia-Pacific, "and it's a costly and less-than-optimal solution for Electrolux. Whenever possible, we will try to work with suppliers to fulfill our requirements."[8]

In a practice it calls "responsible lobbying," Electrolux is also cooperating with governments to develop environmental legislation. A good example is the European Union's Waste Electrical and Electronic Equipment (WEEE) Directive, which went into force in many European countries in 2005.[9] This directive addresses the growing volume of waste in Europe—which includes an additional 50 million or so discarded large home appliances each year. The directive requires the collection and recovery of 80 percent of the weight of large appliances: at least 75 percent must be recycled (5 percent can be converted into energy).

Electrolux claims it "fought strongly" for years, along with nongovernmental organizations like the European Environmental Bureau and the WWF (World Wildlife Fund / World Wide Fund for Nature), to establish the WEEE Directive to make producers responsible for waste. This was necessary, Electrolux argued, to create sufficient corporate incentives to develop products that are easier and cheaper to recycle, such as the hydrocarbons (HCs) it now uses in its refrigerators and freezers: not only are these easier and cheaper to recycle, but they also have less impact on climate change than many alternatives. Making firms responsible for waste also allows a firm like Electrolux to adjust consumer prices to recoup the extra costs arising from recycling and disposal.[10]

Electrolux is involved in a host of other environmental initiatives as well. Along with Braun/Gillette, Hewlett Packard, and Sony, it's a founding member of the European Recycling Platform, the first pan-European effort by industry to manage recycling under the WEEE Directive, covering three-quarters of the total volume of waste in Europe. Electrolux has also taken an active role in developing the legislation to comply with the EU Directive on Restriction of the Use of Certain Hazardous Substances. By restricting six hazardous substances found in electrical and electronic equipment sold after 1 July 2006, this directive required Electrolux to modify nearly all of its electrical products.[11] As part of meeting its commitments here, Electrolux put in place a list of banned and restricted substances for its suppliers to assist with compliance and prepare for future phaseouts. (Electrolux suppliers, such as those in Brazil, work with lower domestic standards, but must still comply with this list when supplying Electrolux for products to be sold in Europe.)

Electrolux is involved, too, in campaigns to educate consumers about the benefits of energy-efficient appliances. Thus, in 2004–05, it released 80,000 copies of an information package in Italy called "Ecoguida" (made in cooperation with the WWF) that shows consumers how to use appliances more efficiently and how to choose an eco-efficient model when buying a new appliance. And, finally, Electrolux is a member of the United Nation's Global Compact, whose corporations have pledged to abide by the compact's ten guiding principles on human rights, labor standards, environment, and corruption—as is the German appliance company Bosch und Siemens Hausgeräte (BSH).

Environmental Responsibility at BSH

Founded in 1967 and best known for its brand name, "Bosch and Siemens," BSH has facilities in some 40 countries with over 35,000 employees. Like Electrolux, BSH publishes annual sustainability reports. "Responsibility for the environment and society is for us an ethical obligation," one report begins, "and at the same time an essential prerequisite for sustainable business success."[12]

BSH aims to set the environmental "benchmark" for home appliances. As a member of the European Committee of Manufacturers of Domestic Equipment (CECED), the company played a leading role in developing CECED's 2005 voluntary code of conduct for corporate social responsibility.[13] BSH's own environmental policy focuses on preventing ozone depletion and climate change—on producing ever more "low-consumption" appliances with ever more efficient use of inputs. This policy has yielded measurable results: from 2002 to 2004, the percentage of BSH refrigerators and freezers receiving an EU Class A rating went from 70 to 85 and from 48 to 64, respectively. The average energy consumption of its refrigerators fell an impressive 78 percent from 1990 to 2004, primarily because of advances in compressor technologies. These achievements are all the more significant given that between 80 and 90 percent of the environmental impact of home appliances occurs during the usage stage.[14]

BSH is also working to use less energy and water and to generate less waste and carbon dioxide during production, packaging, and transportation. Through dialogue with its wholesalers, retailers, consumers, and disposal firms, the company is striving to maximize environmental efficiencies throughout the product life cycle. As with Electrolux, the BSH environmental guidelines require all production sites "where environmental matters are an issue" to receive certification by the International Organization for Standardization under the ISO 14001 standard. Already, by the close of 2004, the ISO had certified 96 percent of these sites.[15]

For both Electrolux and BSH these environmental strategies are part of a broader strategy to retain and capture markets. This is true, too, for the U.S. appliance company Whirlpool, which, as its move to acquire Maytag in 2005–06 shows, is aggressively expanding its operations.

Superefficient Whirlpool

The world's largest manufacturer of "major home appliances," with sales in over 170 countries, Whirlpool has more than 68,000 employees and nearly 50 manufacturing and technology units worldwide.[16] Like Electrolux and BSH, Whirlpool has developed a corporate policy for social and environmental responsibility. "Equal to protecting the health and safety of our employees," the company emphasizes, "we consider environmental stewardship among our most important business responsibilities."[17]

Unlike Electrolux and BSH, however, as of March 2008, Whirlpool was not a member of the Global Compact, although the corporation stresses its commitment to improving environmental performance and protecting ecosystems. In 1993, Whirlpool's Super Efficient Refrigerator Program beat out over 500 firms and inventors for a $30 million winner-take-all contest—a prize awarded by a coalition of 25 public and private electric utilities for designing and mass-producing a CFC-free refrigerator at least 25 percent more energy efficient than U.S. minimum standards. More recently, Whirlpool proudly declared itself the first major appliance company to set a target for reducing greenhouse gases (3 percent below 1998 levels). Its primary means for achieving this target is through developing and manufacturing more energy-efficient appliances; to help market them, Whirlpool has participated in the U.S. government's ENERGY STAR program since 1998.

ENERGY STAR

Established in 1992 as a voluntary partnership between industry and government, ENERGY STAR assists American households and builders with energy conservation, advises businesses on how to produce goods with less energy, and provides consumers with information comparing the energy efficiency of various products (including new homes). The primary means is through the ENERGY STAR label, which provides consumers and businesses with a "credible, objective source of information . . . to make well-informed energy decisions;" products bearing the label must meet energy efficiency standards set by the Environmental Protection Agency and the Department of Energy. Consumers have bought over 2 billion products with an ENERGY STAR label since 1992; currently, more than 1,500 manufacturers use this label on over 35,000 models.[18]

The ENERGY STAR program is designed, not to reduce the consumption of, say, household appliances like refrigerators, but rather to help manufacturers and consumers save energy when producing and using them (and in the process save money all around). According to ENERGY STAR, a typical household that follows its advice can save around a third on its energy bill "without sacrificing features, style or comfort."[19] Thus it becomes possible to lower greenhouse gas emissions by appealing to the self-interest of consumers. The U.S. government sees this as a profitable win-win solution: one that in recent years is gaining power as energy prices rise.

Credible evidence exists of the value of this program. With the assistance of ENERGY STAR, Americans saved $14 billion on utility bills in 2006, averting greenhouse gas emissions equal to running 25 million automobiles for a year (representing about one-third of the total reductions arising from the EPA's climate change programs). This translated into a drop in U.S. demand for electricity of about 4 percent. Over time, the program appears to be getting more effective. The energy savings more than doubled in the five years after 2000. And ENERGY STAR expects these to nearly double again over the next decade.[20]

The ENERGY STAR program is influencing the manufacturing and marketing strategies of multinational appliance companies, too. Whirlpool, for example, has sold over 300 models across 7 product categories that qualify for the ENERGY STAR program. It now has more appliances with an ENERGY STAR rating than any other manufacturer. One example is its Conquest refrigerator, which consumes roughly the same amount of energy per year as a continuously burning 75-watt lightbulb, exceeding minimum U.S. federal energy efficiency standards by 15 percent.[21]

Whirlpool received ENERGY STAR Awards every year from 1999 to 2002 and from 2004 to 2007 (these are given by the Environmental

Protection Agency and the Department of Energy). Whirlpool's 2007 award was for "sustained excellence." The selection criteria took into account efforts to use energy-efficient technologies, to explain the benefits of saving energy to consumers and businesses, and to encourage other firms to join the ENERGY STAR program. Other 2007 winners included some of the world's best-known firms, such as Toyota, Ford, PepsiCo, McDonald's, Home Depot, and Marriott.[22]

Firms and consumers could do many other small things to improve efficiency further. One easy one is to unplug (or design) products to avoid drawing power 24 hours a day. As much as one-fifth of the electricity appliances use occurs in "standby" mode; every year, this consumes $3.5 billion in electricity in the United States alone.[23] Still, the progress in developing better appliances is impressive. Newer models no longer emit CFCs, use less electricity, and are easier to recycle. Average energy efficiency of refrigerators in the United States, for example, has much improved—over 150 percent from 1980 to 2002 alone. By replacing older refrigerators with newer, more energy-efficient ones, consumers can save considerable energy and significantly reduce greenhouse gas emissions per unit. (Deteriorating parts can reduce the energy efficiency of these aging refrigerators by as much as 40 to 60 percent, and simple repairs generally cannot restore their original efficiency.)[24] Yet, at the same time, these significant efficiency gains are part of a process of change that is expanding markets and spurring consumer spending, adding further to the pressures on natural resources and waste sinks. A glance at global trends for electricity consumption is revealing.

Plugging in Global Electricity

Worldwide, consumption of energy has been rising on average by 2.2 percent per year since 1970. Household appliances are the fastest growing drain on energy reserves, after automobiles. Rising sales of home appliances, for example, were the main reason household consumption of electricity more than tripled in China during the 1990s. Household consumption of electricity was growing in other developing countries in the Asia-Pacific over this time as well. The annual rate of increase was 11 percent in South Korea, 13 percent in Indonesia, 25 percent in Thailand, and 28 percent in the Philippines.[25]

Trends suggest even higher consumption of electricity in the future. The U.S. Energy Information Administration predicts a steady annual rise of nearly 2 percent in the next two decades (for a total increase of 57 percent from 2002 to 2025).[26] Already, the main drivers of rising consumption are in emerging economies like Brazil, China, India, Indonesia, Mexico, and Russia—where per capita energy consumption is still well behind the wealthy world. Each year, the United States consumes the equivalent of some 8,600 kilograms (63.6 barrels) of oil per capita. This compares to the equivalent of just over 1,000 kilograms (7.4 barrels) of oil per capita in China and just below 350 kilograms (2.6 barrels) of oil per capita in India. And the United States is not even the highest per capita energy user: in cold places like Canada and hot ones like Singapore, annual energy consumption is the equivalent of well over 10,000 kilograms (74 barrels) of oil per capita.[27]

Fossil fuels are expected to account for some 90 percent of the increase in electricity consumption over the next two decades. The International Energy Agency predicts global demand for oil will increase from about 85 million barrels per day to 116 million barrels per day by 2030, adding further to global carbon dioxide emissions (and thus to warmer temperatures).[28] Changing electricity needs will alter the relative impact of advanced and emerging economies on climate change. This is already beginning to occur. Carbon dioxide emissions rose 15 percent in the United States from 1990 to 2001. The United States, with less than 5 percent of the world population, accounted for 24 percent of global emissions by the end of this period. On the other hand, the amount of carbon dioxide from China went up 35 percent over this time. By 2001, China, with around 20 percent of the global population, was the second largest emitter (accounting for 12.7 percent).

That year, however, China was still far behind the United States in carbon dioxide emissions. Yet, in 2006, just five years later, China passed the United States to become the world's largest emitter of carbon dioxide, in large part because rapidly rising profits for investors and traders in China's liberalizing economy had led to rapid increases both in the construction of coal-fired power plants and in the production of cement, automobiles, and other manufactured goods.[29] Globalization means that a new refrigerator in China is now far superior to an old CFC one from the 1980s. But it also means that increased consumption is casting longer and deeper ecological shadows.

14

The Globalization of Plugging In

Thomas Midgley was no mad scientist. His aims in 1921 and 1928 were reasonable: to get rid of engine knock and to make a safer refrigerator. There was no malice, no cunning plot to set off tragic ecological consequences. He was merely a scientist of his times, dying in 1944 surely far more concerned about how the contraption he'd designed was strangling him than about the safety of Freon.

The scientific consensus was firm for three decades after his death: CFCs were totally safe, stable and harmless wonder chemicals able to cool refrigerators, propel aerosols, and make foams. This consensus held on its own, not because industry had controlled research. Science, quite simply, got it wrong, in large part because, at least until the early 1970s, no one had thought to measure any changes—and thus collect evidence—in the ozone layer so high above the earth.

Mario Molina and Sherwood Rowland's 1974 theory of how CFCs *could* deplete ozone shows the power of science to get it right, too. Their theory would in time help to solve the accidental consequences of Midgley's discovery of the cooling properties of CFCs. But it was only a first step. Many other forces had to come into play, including new political institutions, corporate strategies, trade patterns, financial incentives, and technological advances. Decades would need to pass as well. It would take a decade to overcome companies like DuPont challenging the theory. It would take three decades to negotiate and implement a global phaseout of CFCs. And it will take yet another five decades or so for the ozone layer to recover (most scientists now predict it will return to 1980 levels sometime between 2030 and 2070).[1]

Still, once the Montreal Protocol went into force in 1989, the phaseout of CFC refrigerators occurred reasonably quickly. Governments put in place effective labeling programs as well as mandatory reductions and phaseout schedules. Firms found cost-effective substitutes and retooled manufacturing plants. And consumers upgraded to new models (sometimes with government or corporate incentives). Results came within a half decade. Most notably, by the beginning of 1996 firms were no longer manufacturing CFC refrigerators in the developed world.

This still left a small number of manufacturers of CFC refrigerators in developing countries. In 1990, under an amendment to the Montreal Protocol, these countries agreed to phase out CFC consumption by 2010. Many refrigerator makers, with funding from the Multilateral Fund for the Implementation of the Montreal Protocol, worked over the next decade and a half to reduce CFC use. Some, such as those in China, were converting to CFC substitutes even without international assistance and well before government target times. The main reasons were straightforward. Affordable and reliable CFC substitutes were readily available (sometimes by entering joint ventures with overseas firms or foreign investors). And producing CFC refrigerators gave these firms both access to overseas markets and advantages in the domestic market. Thus, by the beginning of the twenty-first century, CFC substitutes were in place in most of the world's refrigeration systems, with most developing countries now expected to meet the 2010 target comfortably.[2]

Why was the international community able to eliminate the shadow effects of CFCs? Why, more specifically, did corporations and states support a phaseout of CFC refrigerators after the late 1980s? One reason was the solid scientific consensus that emerged in the 1970s: CFCs were depleting the ozone layer—and less ozone meant more skin cancer. Another was the globalization of environmentalism. States were more willing to engage in global environmental negotiations and state environmental agencies were gaining more influence. International aid—especially the Multilateral Fund for the Implementation of the Montreal Protocol—helped bring developing states on board. Growing numbers of multinational corporations were developing codes of conduct to promote environmental and social responsibility. But, as chapters 11–13 have shown, the main reason for this support was the globalization of CFC-free refrigerators. This powerful process first countered, then virtually eliminated resistance to a global phaseout from firms still producing CFC refrigerators, which, unwilling—or unable—to compete in CFC-free markets, went bankrupt.

The globalization of CFC-free markets eased the shadow effects of refrigerators on the ozone layer. At the same time, the world's biggest appliance corporations—ones like Electrolux, BSH, and Whirlpool—have been competing for these markets by implementing codes of conduct and sustainability policies to increase recycling rates as well as reduce waste, pollution, and inefficiencies (by, for example, upgrading factories and modifying packaging). Some companies, as part of their corporate responsibility policies, have also been cooperating more with governments and nongovernmental organizations, supporting, for example, legislation in the European Union to make producers responsible for recycling refrigerators. Some, such as Electrolux, have even been auditing suppliers in places like China and Brazil to improve compliance with corporate codes of conduct. As a result, many of these companies are now meeting, and in some cases even exceeding, international and national standards for better environmental performance (such as ISO 14001 and the 1990 U.S. Clean Air Act amendments.) Many are participating as well in government programs like ENERGY STAR.

Compared to a few decades back, a new refrigerator today is not only CFC free, but also uses less energy and fewer resources during its life cycle. The average energy consumption of BSH refrigerators, for example, dropped nearly 80 percent from 1990 to 2004. Across the United States, the average energy efficiency of refrigerators improved by over 150 percent from 1980 to 2002. Thus an average 1985 side-by-side refrigerator (22 cubic feet) consumed 1,314 kilowatt-hours of power per year at a cost of $112. A 2001 refrigerator of the same capacity meeting ENERGY STAR standards for that year consumed just 576 kilowatt-hours per year at a cost of $49.[3]

Nevertheless, over this same period, the impact of the refrigerator sector on the global environment has been spreading and intensifying as rising consumption overrides the gains in producing, using, and disposing of each new unit. Consumption is rising per capita in mature markets like the United States and Europe, as firms encourage consumers to make quick upgrades, purchase a second or third refrigerator, and buy larger models. Consumption is rising even faster overall as appliance companies like Electrolux, BSH, and Whirlpool move into developing countries. This is contributing to rising household consumption of electricity across the developing world—consumption more than tripled in China in the 1990s alone. Thus, even as the ecological shadow of refrigerators moves away from the ozone layer, it grows deeper in the developed world and longer in the developing world, with the ever growing consumption of refrigerators using up more natural resources, generating more waste, and producing more greenhouse gases.

Rising global consumption over the last century—as in the cases of the automobile and the refrigerator—is the norm for most manufactured goods. It's also the norm for natural resources like water, timber, and land, which, as the next case will show, is in part a result of rising consumption of foods like beef.

IV

Beef

15

The Efficient Steer: Fast, Fat, and Cheap

In *An Essay on the Principle of Population* (1798), the scholar Thomas Malthus put forth a seemingly inevitable principle: population, left unchecked, increases exponentially, while food production increases only arithmetically. Thus, following the laws of mathematics, mass starvation must one day ensue, causing a die-off of the human race.

Yet Malthus was wrong, at least about the second premise of his principle: the production of food over the last two centuries has been able to keep up with—and often surpass—the exponential growth of the human population. Today, there's more than enough food for the world's 6.7 billion people, and most starvation arises, not from a basic lack of food, but from inadequate distribution, incompetent governments, or overconsumption.

Malthus's fundamental error was underestimating the capacity of technology to increase efficiencies, extend productive land, and deflect costs into remote ecosystems and the future. He did not foresee the potential of irrigation, pest-resistant seeds, chemical fertilizers, and pesticides to boost crop yields. Nor, for foods like beef, did he foresee the capacity of businesses to produce a more efficient steer by injecting it with hormones to grow faster, feeding it buckets of corn and soy to fatten it more quickly, confining it in a feedlot to keep it marbled and heavy, and treating it with antibiotics to help it survive these unnatural conditions. He did not foresee the potential of disassembling that steer on a fast-moving conveyer belt in industrial slaughterhouses, paying workers next to nothing to carve out every scrap, and of then transporting its meat—ground, dried, frozen, canned—from faraway farmlands to cities. And he did not foresee how efficiently and cheaply multinational corporations could distribute these meat products through supermarket chains and fast-food outlets to the hungry (and not-so-hungry) masses. How, after all, could he have imagined such a future?

Industrial farming, as the history of commercial beef in this chapter shows, has saved many humans from starving. Indeed, in many countries, overconsumption of beef, among other foods, has made obesity a far greater threat to human health than starvation. But rising obesity is not the only unintended consequence of the much expanded production and consumption of beef. As this chapter argues, keeping the output of beef apace with rising populations and surging per capita consumption has generated many shadow effects for people and ecosystems (including farm animals).

Inside the Slaughterhouses

A Lithuanian immigrant from the imagination of Upton Sinclair, Jurgis Rudkus lives a life of unremitting misery. He toils in a slaughterhouse of Packingtown, Chicago, at the beginning of the twentieth century. Shrouded in gloom, knives slash at carcasses with lightning strokes. The floors and walls are cold, slippery, and bloody, the stench gut wrenching. The bosses are brutal, whipping workers to make the *dis*assembly line go faster. The workers are given no training, no benefits, and virtually no wages. Driven by greed, the meatpacking oligopoly greases a city of graft run by gangster politicians.

Every scrap of meat—even the rancid and disease ridden—is shoveled, along with rats and their feces, into the ground beef and sausages, canned and pickled as delicacies. Before long, Jurgis is cheated into debt, then injured and cast aside, without work or pay. Forced to struggle even harder to support the family in this slaughterhouse hell, his young bride goes mad after being sexually abused by her boss.

Then, having served jail time for beating his wife's abuser half to death, when it seems life couldn't possibly get any worse, Jurgis is unable to scrounge up enough money to save his wife from dying in childbirth. Grief stricken, he slogs on, working at menial jobs in Chicago to support his toddler son. When, however, the boy drowns in a ditch steps away from his ramshackle home, life loses all meaning for Jurgis. After this, he lurches about without purpose, becoming a hobo, a strikebreaking scab, a political lackey, a falling-down drunk.

Sinclair's moralizing novel, which without any subtlety he titled *The Jungle*, ends after Jurgis discovers socialism, and from the ashes of his despair comes the hope of an electoral uprising of the workers of America. In his account of a Chicago slaughterhouse in the early years of industrial ranching, Sinclair had sought to show how the pursuit of efficiencies and savings was turning the family farm into a factory, with bosses exploiting workers to produce more "food," faster and cheaper. *The Jungle* is arguably Sinclair's most influential work in a lifetime of publishing over 90 books. But the reaction of the American public disappointed him. "I aimed at the public's heart," he later wrote in his autobiography, "and by accident I hit it in the stomach."[1]

The reaction of American consumers to such stories—disgust and outrage, then demands for government controls to ensure sanitary working conditions and the quality of meat—has been common elsewhere, too. Meat sales have sometimes declined after readers experienced the shock of "seeing inside" a slaughterhouse, but this has always been temporary, and, as this chapter documents, per capita meat consumption has been rising in every culture since the early 1900s.

Producing Pure Food

The Jungle was a bestseller in the United States. But it did not cause the mass uprising Sinclair had hoped for. To his dismay, the outrage stirred up by his novel was not over the hardships of Jurgis and his fellow workers, but over the unsanitary slaughtering in the meatpacking plants. Wasn't the health of consumers at risk? Sinclair had spent many weeks researching his novel in the meat slums of Chicago: his description was vivid, specific, as seemingly real as any journalistic exposé. Beef sales began to tumble as sales of his novel climbed. Public pressure for action grew; before long, the U.S. Congress passed the Pure Food and Drug Act and the Meat Inspection Act in 1906.[2]

These acts, which created the U.S. Food and Drug Administration, gave the government better control over the quality of the meat bought by the average American consumer. Conditions for U.S. meatpackers in the first half of the twentieth century began to improve, too, under pressure from public health advocates and unions. The stomach-turning scenes in *The Jungle* did not, however, produce a manifesto for a world vegetarian revolution, and after the initial drop in U.S. beef sales, did nothing to alter the trend toward eating more meat. Instead, as this chapter also documents, over the last century, the capitalists of Sinclair's world—who were using assembly lines even before Henry Ford—have managed to further improve the "efficiencies" of producing beef.[3] Ranches are bigger. The cattle fatten faster on a diet of cheap grain, growth hormones, and antibiotics. High-tech disassembly plants process the beef. And cattle graze in chemically fertilized pastures and cleared rainforests.

As a result, industrial farmers over the last century have been able to produce enough beef to easily outpace the needs of growing human populations, so much so that many people now consume too much for a healthy diet. The trend toward consuming ever more beef, as chapter 16 will document, is an increasing strain on environmental resources, from local waterways to tropical rainforests to the global climate. On a more optimistic note, chapter 17 will chart a shift among some consumers toward eating more "sustainable" beef, such as natural, organic, or grass fed. Yet, as that chapter will show, such environmentally friendly change is chasing a stampede of demand for cheap steaks and ground beef sold by industrial meatpackers. Understanding the reasons for this demand requires us to step back and look at the history of consuming farmed meat.

Farming Meat

Agricultural societies began to emerge 8,000 to 10,000 years ago, when nomadic hunter-gatherers began to settle in fixed locales. The resulting increase in consistent food supplies spurred a trend toward larger towns and, eventually, cities. Still, farming practices in traditional agricultural societies did not allow for quick weight gain in domesticated animals. Most farmers kept cattle, horses, or camels for transportation, for plowing and dunging their fields, and for producing milk, rather than for meat. Indeed, archaeological evidence and written records suggest that per capita meat consumption was generally low and stable in most traditional agricultural societies—rarely more than 5–10 kilograms (about 10–20 pounds) a year.[4]

Peasants in many of the subsistence societies of Europe, imperfect records suggest, rarely ate meat more than once a week, and large quantities only during celebratory feasts. Although nobles, wealthy landowners, marching armies, and city dwellers tended to eat more meat than peasants did, their numbers were comparatively small. Animals generally supplied less than 15 percent of all dietary protein in Europe, even into the eighteenth and nineteenth centuries. According to one study, meat accounted for less than 3 percent of the average annual food energy in early-nineteenth-century France. Another study calculated the per capita meat consumption of poor Welsh and English laborers in the late 1700s at just over 8 kilograms (18 pounds) a year. Still another put the annual average consumption of meat in Germany in 1820 at less than 20 kilograms (44 pounds). The per capita consumption of meat was even lower in most other parts of the world, such as in China, India, and Japan, although colonial settlers in countries like Argentina, Australia, and New Zealand were beginning, even before industrialization, to consume much higher amounts of meat (especially beef and mutton) than those consumed in Europe.[5]

Eating habits changed significantly in western Europe and North America after the mid-1800s as agricultural output rose, cities expanded, and industrialization intensified. The average diet began to include more meat, fish, dairy foods, fruit, and sugar—and less staple cereals and legumes; rising imports of foods from colonies also provided more choices. The beef industry expanded particularly quickly. Beef producers in countries like the United States began to integrate small ranches into industrial meatpacking plants. In 1850, just 185 meatpacking plants were operating in the United States, producing $12 million worth of red meat; by 1919, there were over 1,300 plants, producing $4.2 billion worth.

Similar shifts in diets and meat processing began to occur across the globe over the next 100 years. After World War II, the pace of change accelerated with new crop varieties, new chemical sprays and fertilizers, ever larger mechanized farms, and more efficient processing techniques, and as these changes spread to the populous countries of the developing world, most notably those of East Asia.[6]

Rising Consumption of Meat

The number of farm animals has been climbing rapidly since the 1950s. There are now over 1 billion pigs, 1.3 billion cattle, 1.8 billion sheep and goats, and 17 billion chickens. Worldwide, annual meat production has jumped more than fivefold since 1950—to over 260 million metric tons. Annual per capita consumption of meat from 1950 to 2005 more than doubled: from 17 to 40 kilograms (38 to 88 pounds). Beef accounts for around 25 percent of this total, behind pork at 38 percent and poultry at 30 percent.[7]

China is by far the world's largest national consumer of meat, with annual consumption now over 68 million metric tons and rising, in large part because of steadily increasing per capita consumption, at just over 52 kilograms (115 pounds) a year in 2002. China consumes far more meat than other heavily populated countries like India and Indonesia. India consumed 5.5 million metric tons of meat in 2002, and Indonesia 1.8 million metric tons; this translates into an annual per capita consumption for that year of 5 kilograms (11 pounds) in India and 8 kilograms (18 pounds) in Indonesia. China is increasingly relying on grains and soy meal to sustain its livestock (and promote rapid weight gains); by 2000 it was already using about one-quarter of its grain to feed livestock—twice as much as in 1980.

Although the United States is the world's second largest national meat consumer, with annual consumption now over 39 million metric tons, its per capita consumption—at 125 kilograms (275 pounds) a year in 2002—is far higher than China's.[8] Beef remains at the core of the American meat diet. The consumption of beef in the United States took off after 1870: the cattle shipped over by European colonists were thriving in the open plains of the American West and the market for beef was expanding as refrigerated railway cars allowed more beef to reach consumers in the growing cities of the East Coast.

Americans, on average, were eating 23 kilograms (about 50 pounds) of beef a year in 1910–15. This average would rise and fall over the twentieth century—from a low of just under 19 kilograms (about 40 pounds) in 1930–35 to a high of close to 39 kilograms (about 85 pounds) in 1970–75—with an overall annual average for the twentieth century of nearly 27 kilograms (about 60 pounds) per person. By the start of the twenty-first century, average annual consumption of beef had risen to around 29 kilograms (64 pounds) per person, an amount not dramatically higher than in 1909 (when the U.S. government first began to keep records) and one that has held fairly steady despite regular advertising campaigns to encourage more beef consumption.[9] The reason is simple: Americans began to eat far more poultry—from an annual total of less than 5 kilograms (about 10 pounds) per person in 1909 to nearly 27 kilograms (about 60 pounds) in 2004. Largely as a result, beef as share of total meat consumption in the United States has declined over the last 100 years, from around 45 percent in 1909 to less than 35 percent in 2004.[10]

Some cultures, such as those of India, have long traditions of vegetarian diets. Others, such as many in Asia, have long culinary histories of popular dishes with little or no meat. Various surveys of consumers in countries like the United States and United Kingdom have also shown rising interest in vegetarian choices. Still, the percentage of vegetarians remains low in wealthy Western countries—with surveys usually finding that between 4 and 10 percent of respondents identify themselves as "vegetarian" (in various senses and with various degrees of commitment).

Although more people across many cultures seem to be choosing a vegetarian diet, this is having little statistical impact on the worldwide consumption of meat as the human population rises and as increasing numbers of people eat more meat. Average meat consumption in developing countries, for example, was 10 kilograms (22 pounds) per person in 1964–66; by 1997–99, it was 26 kilograms (57 pounds). The Food and Agriculture Organization (FAO) expects it will rise to 37 kilograms (82 pounds) per person by 2030—despite continuing rapid rates of population growth.[11] The globalization of industrial meat production over the last century explains how the world has been able to supply so much meat to so many people. The beef industry is typical—with U.S. meatpackers, agricultural companies, and fast-food corporations playing leading roles.

The Beef Industry

Many consumers still imagine beef comes from a vast and rugged ranch—from Wild West Texas or the Aussie outback. Indeed, in much of the world, this is the case. But hundreds of millions of cattle also live at least part of the year in crowded and confined feedlots. Such factory farming methods now account, according to some estimates, for over 40 percent of global beef production.[12] The animals on many industrial farms live with little natural light or fresh air. To "produce" veal, for example, some farms separate calves from their mothers a few days after birth, lock them in stalls so small they cannot lie down or groom comfortably, then feed them a liquid diet from buckets to keep the "meat" tender and pale to meet consumer preferences. These calves are typically slaughtered after 16 weeks.

Few consumers have seen such stomach-turning practices, and fewer still have openly protested the treatment of cattle (unlike the many who have protested the hunting of whales or harp seals). The total number of cattle in these conditions has been rising steadily over the last 100 years. Some 500 million head of cattle roamed the earth at the start of the twentieth century. Now there are nearly three times as many.[13] But the near tripling in the total number of cattle does not accurately reflect an even bigger rise in the consumption of beef as producers bring heavier cattle to the market even faster than when the "Beef Trust," an oligopoly of wealthy Chicago meatpackers, controlled the U.S. beef industry at the start of the twentieth century.

The American Beef Industry

The power of the Beef Trust in the United States peaked in 1917, when the five biggest meatpacking firms accounted for over half of the market. Then, under pressure from "trustbusters" in the federal government, and following a Federal Trade Commission inquiry into collusion among firms to divide markets and fix prices, the largest meatpacking firms agreed to sign a consent decree in 1920 that forced them to sell off stockyards, retail stores, railways, and livestock journals.

The following year, Congress established the Packers and Stockyards Administration to combat price-fixing and collusion in the beef industry. Over the next half century, small ranchers received better prices for cattle through open bidding at auctions. The working conditions within many meatpacking plants were improving as unions won increases in wages and benefits and government regulators forced higher standards for safety and sanitation. These improvements were not to last, however. In the 1960s, the Iowa Beef Packers (IBP) began to recruit migrant workers from Mexico for plants in rural areas (away from union strongholds). As other meatpackers followed suit, wages across the whole industry fell markedly over the next two decades.

Regulation of the beef industry took a sharp turn during the administration of Ronald Reagan (1981–89). By 1980, the market control of the largest beef producers was far less than in the days of the Beef Trust. The Reagan administration, however, began to allow meatpacking firms to merge and gain control over local cattle markets.[14] Today, just four firms control more than 80 percent of the meatpacking: Tyson Foods (which acquired IBP in 2001), Excel (a subsidiary of Cargill), Swift and Company (formerly ConAgra Beef), and the National Beef Packing Company. This market control, as bestselling author Eric Schlosser writes, "is now at the highest level since record-keeping began in the early 20th century."[15]

Today, for many workers, meatpacking is again, as in the days of Jurgis Rudkus, a low-paying and dangerous job, even as U.S. consumers spend about $70 billion a year on beef. A typical plant, according to Schlosser, "now hires an entirely new workforce every year or so."[16] Many of these workers are illegal immigrants.

Because of different geographies, farming traditions, and regulatory systems, considerable differences exist in how ranchers and meatpackers treat cattle across and within countries. Some American farmers, for example, rely solely on grass to feed cattle. Most cattle, however, eat grass for six months or so on a ranch, then, in the language of industrial farming, are "finished" during the fall and winter months in feedlots holding as many as 100,000 cattle, where they feed on grains, often mixed with antibiotics and protein supplements, to fatten them as quickly as possible.[17]

Corn is the most popular ingredient for cattle feed in the United States, with government subsidies ensuring cheap and plentiful supplies; 50–60 percent of the corn harvest in the United States is now fed to livestock. Grazing cattle, on average, gain no more than 0.5 kilogram (1 pound) a day, whereas cattle in feedlots tend to gain more than twice as much, over 1 kilogram (2 pounds) a day. Many cattle also receive growth hormones, which can increase weight gain by 20 percent. Over 90 percent of cattle raised in the United States by industrial methods now receive growth hormones through injections or implants (a practice banned by the European Community in 1988).[18]

Together, these practices in the United States result in a highly efficient weight-gain program for cattle. Steers at the beginning of the twentieth century would commonly live at least 4–5 years before slaughter. By the 1950s, ranchers were able to get a steer to slaughter weight within 2–3 years. Today, the new antibiotic feeds and hormones can enable a calf to gain over 500 kilograms (and reach a profitable slaughter weight) in as little as 14 months.[19]

Ranchers commonly truck these cattle to slaughterhouses able to process hundreds of carcasses every hour.[20] As in the days of Jurgis Rudkus, efficient slaughtering still relies on workers with razor-sharp knives along a disassembly line. But meatpacking firms have found additional efficiencies and savings, too. Today's machines can, for example, slice even more "legal meat" from carcasses. The merger of firms during the Reagan administration substantially reduced operating costs. Busting unions and relying on illegal labor have made operations cheaper still. Disassembly lines are now faster, too, with some able to handle close to 400 cattle per hour, almost twice the typical rate of 25 years ago. Giant slaughterhouses, such as those of Tyson, Excel, Swift, and National Beef, have been able to reduce costs by as much as 40 percent since the early 1980s, with the result that, according to the Department of Agriculture, wholesale beef prices have gone down almost every year since then.[21]

On the other hand, these advances in producing more affordable beef have not been without costs. The "advanced meat recovery systems" of the 1990s relied on hydraulic pressure to strip any remaining meat off the processed carcass bones. This extra meat was valuable filler for hamburgers, hotdogs, and pizza toppings. But applying too much pressure—or removing the spinal cord improperly—laced meat with bone and nerve tissue: the USDA, for example, found spinal cord tissue in some of the meat in 1997. Consumer groups, such as the National Consumers League, called for a ban on such meat recovery systems, arguing that the meat recovered was, not proper beef, but beef-bone mush that was leaving consumers at greater risk of mad cow disease.

At the time, the beef industry saw little reason to panic. Used properly, recovery systems did not taint beef. Besides, unlike cattle in the United Kingdom, cattle in the United States did not suffer from mad cow disease. Still, under pressure from consumer groups, and worried about public reactions, some major buyers—notably General Mills and McDonald's—decided to no longer use advanced recovery beef. Several meatpacking firms—facing stricter regulations, rising costs of supervising advanced recovery machines, and purchasing policies of firms like McDonald's—decided to mothball these machines. By 2004, the number of processors using them had fallen from 35 to below 30.

For ranchers, the capacity to fatten and process cattle more quickly has not necessarily meant greater profits. The beef industry has always been a tough business. According to the publisher of the *Cattle Buyers Weekly*, relatively high labor costs and variable cattle prices mean that profit margins for meatpacking firms rarely exceed 2 percent.[22] Dependent on these firms, the latest generation of ranchers raising cattle during the spring and summer months for sale to feedlots are facing especially hard times. "Hell," a South Dakota rancher grumbled in a 2002 interview, "my dad made more money on 250 head than we do on 850."[23] Slim profit margins, however, do not mean meatpacking firms are not prospering. Just the opposite: over the last half century, the globalization of industrial beef production has played a central role in the growing profits of the fast-food industry, whose continued expansion is generating even more demand for cheap beef.

Overconsuming More, Faster

The growth of the fast-food industry over the last 50 years has altered patterns of meat consumption across many cultures. With restaurants in over 120 countries and territories and with record-high revenues in 2007 of nearly $23 billion, the world's largest fast-food chain, McDonald's, serves over 50 million customers a day and is now the largest buyer of beef in countries like the United States. Many other fast-food chains featuring hamburgers, such as Burger King, Wendy's, A&W, and Hardees, are serving many millions more customers. Still other fast-food outlets offer different choices—fresh submarine sandwiches, thick-crust pizza, fried chicken, and spicy tacos, for example. Indeed, the world's largest submarine sandwich chain, Subway, now operates over 26,000 restaurants in over 80 countries—with, it brags, more outlets in the United States, Canada, and Australia than McDonald's. The world's largest chain of pizza restaurants, Pizza Hut, operates in over 100 countries and territories. The world's biggest chain of fried chicken restaurants, Kentucky Fried Chicken, serves 8 million customers a day from more than 11,000 restaurants in more than 80 countries and territories. These last two chains are part of the world's largest restaurant "system," YUM! Brands—a parent company that also owns Taco Bell, A&W, and Long John Silver's, giving it control of more than 35,000 restaurants.[24]

Supersizing meals is a common strategy among these fast-food chains to entice and keep customers. Take Burger King's Stacker: four slabs of beef, four strips of bacon, and four slices of cheese, for a total of some 1,000 calories. The sales pitch to consumers is hardly subtle: in one Burger King commercial in 2006, a manager yells, "More meat!" at workers making a burger. Nor is the Stacker even the biggest hamburger. Hardee's Thickburger, for example, weighs in at over 1,400 calories— about 70 percent of a typical person's recommended daily calorie intake. Advertising fast food as a deal, as getting "more" for "less," is true even for restaurant chains like Subway, which proudly claims that its subs are so healthy and so fat free that eating them every day is an easy way to diet—a sales pitch the company runs alongside its latest advertising jingle: "Double the meat, double the cheese!"

Other fast-food chains, seeing the growth of ones like Subway, have begun to offer "healthier" choices as well, such as salads, fruit bowls, veggie burgers, and bottled water. The big profits, however, remain with the Big Macs, the Whoppers, the Stackers, and the Thickburgers. Some chains have even lost money on the healthy items. Wendy's fresh fruit bowl, for example, failed to sell well even after a $20 million advertising blitz in 2005. "We listened to consumers who said they want to eat fresh fruit," explained a spokesman for the fast-food chain, "but apparently they lied."[25]

Given such trends, the current global crisis of rising obesity is hardly surprising. The United States has some of the world's highest rates of obesity: two-thirds of adults are now overweight (Body Mass Index of 25 or more) or obese (BMI of 30 or more). Children are overweight, too:

currently about one in six children from 6 to 19 years of age. The U.S. Surgeon General now estimates the total medical cost of illnesses related to obesity at well over $100 billion a year. This is a sharp rise from just two decades ago, when fewer than half of adults were overweight.

Rates of obesity are rising in the rest of the world as well, even in the developing world, where changing lifestyles and diets increasingly high in sugar and fat are causing even the undernourished to gain weight. Over one-and-a-half billion adults are now overweight, with at least 400 million of them obese. Even children under the age of five—some 20 million worldwide—are now overweight.[26]

In this era of Burger King Stackers and Meat Lovers Pizzas, the global consumption of meat is expected to rise even more over the next few decades. On average, people are already consuming around 175 pounds (about 80 kilograms) of meat a year in the developed world. Still, wealthy consumers are capable of eating far more, as annual per capita meat consumption in the United States shows (about 100 pounds higher). Yet the biggest potential for growth is in the developing world. Although, on average, people there are consuming far less meat, per capita consumption has been steadily climbing since the 1970s, and although it will remain well below the First World, all trends suggest the Third World will continue to close the gap over the next decade or so (reaching about 80 pounds per person by 2020).[27] The beef industry will supply much of the future worldwide demand for meat—a trend, as we'll see in chapter 16, that will further intensify and extend the ecological impacts of consuming beef.

16

The Ecology of Big Beef

Producing so much beef involves many ecological costs. Farmers are tilling land with pesticides and fertilizers to grow enough grain to fatten cattle quickly. Waste from feedlots is polluting local waterways and air. Growth hormones are tainting food chains, and antibiotics are flowing through ecosystems. The nutritional value of beef is inconsistent and declining in some places. Ranches and feed crops like soybeans are deforesting biodiversity hotspots like the Amazon. And grazing, fattening, and slaughtering billions of cattle every few years is depleting water supplies and emitting vast quantities of greenhouse gases like methane, nitrous oxide, and carbon dioxide.

These impacts are intensifying as the globalization of beef markets creates opportunities to expand commercial ranching even further. Plantations for animal feed crops like corn and soybeans are spreading in response. This in turn is contributing to a surplus of cheap vegetable oils, with incentives for firms to get people to consume more, whether as salad or cooking oil, in margarine and processed food, or in bakery shortening. What the health consequences will be is an ongoing experiment.

Much of this growth in industrial ranching and agriculture is occurring in developing countries; the beef and feed grains produced there are then exported primarily to developed or transitional economies. Many of these operations are supported by government subsidies, foreign aid, and by multinational companies, whose "support" is increasing the foreign debt of these developing countries and profiting the multinationals disproportionately. Prices that do not account adequately for the social and environmental costs—most notably, the impacts on water, land, and climate—are also stimulating overconsumption as cheap beef becomes more common in more cultures. This globalization of industrial beef production, this chapter argues, is intensifying the shadow effects of consuming beef, with the costs increasingly deflected into developing countries, the global commons, and people's future health.

Planting Grain-Fed Beef

The increasing global capacity to produce grain during the twentieth century was essential for feeding a rising human population. It also began to alter animal feeding practices. Worldwide, about one-tenth of grains went to feeding farm animals in 1900—mostly for animals working in the fields. This share had risen to one-fifth by 1950. The switch to feeding beef cattle on grain in the First World was spurred along as chemical fertilizers increased grain yields. By the 1960s, developing countries were also beginning to produce grain surpluses, as new seeds, fertilizers, and pesticides of the green revolution allowed crops to grow faster and in harsher conditions. Foreign aid, technical advice from organizations like the Food and Agriculture Organization, and low-interest loans from organizations like the World Bank—along with the investments of companies like Ralston Purina and Cargill—encouraged many developing countries to focus on growing grain for animal feed and even to switch to coarse grains more suitable for livestock.

As a result, the share of grain fed to livestock in the developing world tripled in the second half of the twentieth century (exceeding 21 percent by the end of the century). Over this time, many developing countries also began to export grains to animal-feed markets in developed countries. Countries like Ethiopia, for example, began to produce grain meals to feed livestock in Europe rather than growing food for their own people (with sometimes grave consequences, such as during the 1984 famine in Ethiopia).

These changes occurred even though many industrially produced grains are not natural foods for farm animals. Feeding corn to cattle, for example, can cause bloating, digestive disorders, sometimes even death. Compared to grass feed, corn feed also tends to produce beef higher in fat and lower in omega-3 fatty acids (which have been found to prevent heart disease and to strengthen the immune system). Still, over the last half century, farmers in many countries have been turning increasingly to grains—especially corn—as a cheap and effective way to fatten animals quickly.

Worldwide, over one-third of grain production now goes to feed livestock, with countries like the United States devoting over 60 percent to that purpose, and countries like India less than 5 percent.[1] Feeding beef cattle so much grain is an efficient means of fattening, but it's an inefficient use of environmental resources. On an industrial ranch, it generally takes 11–17 calories of feed to produce a calorie of beef. Typically, this means it takes one-third more fossil-fuel energy to produce a calorie of beef than, say, a calorie of potatoes. It requires far more water as well. Producing a kilogram (2.2 pounds) of beef can require as much as 125,000 liters (33,000 gallons) of water. And even the more average case of 10,000 liters (2,650 gallons) of water to produce a kilogram of beef is far higher than, say, the water it takes to grow a kilogram of rice or wheat.[2]

Grains are not, moreover, the only agricultural crops used to feed domesticated animals. Farmers over the last half century also began to mix increasing amounts of soybean protein meal with grain, primarily because this can nearly double the efficiency of grain to convert into animal protein. A glance at the global soybean industry over the last 50 years reveals how changing agricultural practices are impacting the nutritional characteristics of global food supplies.

Stuffing Meat with Soy

Global production of soybeans was 16 million metric tons in 1950. Since then, the industry has steadily built new markets: by 2005, production had risen to 220 million metric tons—nearly 14 times higher.[3] Surpluses of soybeans in the United States over the last half century partly explain these expanding markets: the United States, for example, exported surplus soybeans into Europe under the Marshall Plan after World War II. American government subsidies for soybean farming have caused some of these surpluses. Even today, these subsidies remain substantial, with the soybean sector receiving $13 billion between 1998 and 2004.[4]

The phenomenal growth of the soybean industry was made possible in the 1940s, after scientists discovered how to deactivate the enzyme inhibitor in soybean meal so that animals could tolerate it as feed. Soybean meal constitutes nearly 80 percent of the crushed beans after the oil is extracted. Today, it's by far the world's largest source of protein feed for chickens, cattle, pigs, and fish—accounting for 65 percent of global supplies. Around 98 percent of soybean meal goes into livestock feeds in countries like the United States.

Indeed, the United States produces and exports more soybeans than any other country does—around 35 percent of the world's total supply, worth some $19 billion per year in recent years; soybeans are second only to corn among U.S. crops (farmers commonly rotate soybean and corn crops). Three American companies—Archer Daniels Midland (ADM), Bunge, and Cargill—dominate the soybean market in the United States. These firms have managed to increase the share of soybean meal in livestock and poultry feed in the United States from less than 10 percent in 1964 to almost 20 percent today.[5]

The influence of these companies, however, extends far beyond the United States. They control nearly 80 percent of Europe's soybean-crushing industry and nearly 80 percent of its animal feed manufacturing. In producing soybean meal for animal feed, the soybean-crushing industry produces oil for industrial processes and human consumption. The production of soybean oil rose rapidly alongside soybean meal from 1965 to 2005, especially after the industry was able to improve the oil's smell and taste—increasing almost sevenfold, from 5 to 34 million metric tons.[6]

Soybeans are currently the world's largest source of vegetable oil.[7] Soy in some form is now found in a wide range of foods, including breakfast cereals, breads, noodles, soups, cheeses, mayonnaises, and sausage casings. Over 60 percent of processed food now contains soy in countries like Britain. The fast-food industry uses hydrogenated soybean oil for deep-frying, too. Soybeans now account for about 90 percent of oilseed production in the United States—with canola, cottonseed, rapeseed, peanuts, and sunflower seed trailing far behind. Indeed, soybean oil, used mainly as cooking and salad oil, in margarine, and in bakery shortening, now accounts for about two-thirds of total U.S. consumption of vegetable oils and animal fats. Few consumers seem to worry about this trend; many see soy as a healthy choice. As tofu, in veggie burgers, and in soy milk, it is the basis of many vegetarian diets. And, in countries like the United States, it constitutes nearly one-fifth of infant formula.[8]

But is soy really healthy? The American soy industry wants consumers to think so: it spends almost $80 million every year to research ways to promote more consumption—research the industry finances from a mandatory levy on producers. A glance at Japan, where soy plays a central culinary role—and where life expectancy, at over 80 years, is one of the world's highest—would seem to suggest that soy is indeed a healthy source of protein.

But, because soy contains toxins and plant estrogens, some researchers are now wondering whether, as with so many foods, too much soy might prove unsafe. Some experiments have linked high soy consumption to thyroid damage and disruptions in menstrual cycles. In 2002, a British expert committee reported on the risks of high soy consumption for some age groups. Still, the soy industry continues to expand; new soybean plantations now reach deep into the rainforests of Brazil. Driving this expansion, the American firms ADM, Bunge, and Cargill now account for some 60 percent of Brazilian soybean exports.[9]

The increase in soy output, then, is changing patterns of global nutrition, directly by flowing into processed, fast, and even health foods, and indirectly by helping farmers to produce more kilograms of beef more cheaply and quickly—and causing them to alter the nature of that beef. Fattening cattle with grain and soybean meal, moreover, requires regular doses of antibiotics to keep the bloated and confined herds "healthy" and fast growing.

Feeding Antibiotics

The feed for the beef industry is flooding ecosystems with antimicrobial drugs, including antibiotics. In the United States, for example, cows, chickens, and pigs receive 50–70 percent of all antimicrobial drugs. Farmers have been adding antimicrobial drugs to livestock feed and water since the 1950s, both to allow animals to gain weight faster on less feed and to prevent illnesses and diseases from spreading, especially in farms with large and homogeneous herds living in tight quarters with poor ventilation. Antibiotics allow ranchers, in the words of one staff veterinarian in Kansas, to "feed" cattle "hard" on corn, soybean meal, and other protein supplements while avoiding a high "death loss" in animals whose delicate digestive systems were designed to convert grass into protein.[10]

Worldwide, half of all antibiotics (by weight) go to livestock and fish in an effort to prevent disease. The use of antibiotics like penicillin, tetracycline, and erythromycin has been rising over time: in the case of beef cattle in the United States, for example, farmers now use at least 28 percent more antimicrobial drugs than in the 1980s. Many of these seem to flow into animal waste undigested—one study found between 25 and 75 percent did—along with bacteria resistant to antibiotics, which can then pose a threat to the health of humans. Some researchers, such as David Wallinga at the Institute for Agriculture and Trade Policy, see antibiotics as a growing danger. "We're sacrificing a future where antibiotics will work for treating sick people," he says, "by squandering them today for animals that are not sick at all."[11]

Heavy doses of antimicrobial drugs don't necessarily make feed safe for cows—or people. For example, bovine spongiform encephalopathy (mad cow disease) is transmitted when bone and other waste tissue from infected carcasses is mixed into cattle feed. Mad cow and other diseases impervious to antimicrobials can spread quickly through vast herds of similar breeds living close to one another—with beef exports and imports multiplying the potential for rapid spread across the globe. Injecting cattle with hormones to stimulate rapid growth also poses health risks to consumers, directly from the beef consumed and indirectly from contamination when farm waste seeps into surrounding water and soil and then into food chains.

Industrial ranching has other consequences for the global environment. The artificial feed leaves the cattle bloated and, without antimicrobial drugs, often sick. Belching and flatulent livestock now account for one-quarter to one-third of worldwide methane emissions from human-related activities. Meanwhile, decomposing manure emits nitrous oxide, which, like methane, is a primary greenhouse gas driving climate change.[12] Most of the energy to raise cattle (growing grain for feed), to process the carcasses (running the slaughterhouses), and to distribute beef (trucking and refrigerating) is generated by burning fossil fuels, which adds still more to global emissions of carbon dioxide. Livestock and livestock waste as a whole contribute to somewhere between 5 and 10 percent of global greenhouse gas emissions. Fattening cattle in feedlots in particular tends to produce large amounts of carbon dioxide—by one analysis, more than twice as much as grazing them on open range- or pastureland. Burning down forests to create pastures in places like the Amazon is also an increasingly large source of carbon dioxide emissions—and an increasingly serious threat to biodiversity.[13]

Ranching the Amazon

The Brazilian Amazon, comprising nine states and covering 500 million hectares (1,930,000 square miles; over 50 percent of Brazil's total land area), holds some of the world's highest concentrations of biodiversity.[14] The Amazon region lost over 17 million hectares (65,500 square miles) of forest in the 1990s alone—an area about the size of Uruguay. By 2000, the total deforested area in the Amazon was nearly 59 million hectares (227,500 square miles).

As more ranchers have cleared more land to graze cattle for a surging beef export market, the average annual deforestation rate has climbed.

Before 1990, ranchers in the Amazon sold most of their beef within this region. The market for Amazon beef became more national in the 1990s: with rising urban incomes, beef consumption in Brazil quadrupled overall (and more than doubled per capita) from 1972 to 1997. Then, in the late 1990s, when low land prices, a devaluation of Brazil's currency, and better control of foot-and-mouth disease made it more profitable, production of beef for export took off. Beef exports grew fivefold from 1997 to 2003—with European Union countries importing close to 40 percent of Brazil's fresh and frozen beef in 1999–2002, followed by Chile and then Egypt.[15]

In 2003, Brazil produced $1.5 billion worth of beef for export—over 3 times more than in 1995. Brazil's beef exports by carcass weight equivalence (which excludes the head, hide, and intestines) grew at an even faster rate—from 232,000 metric tons in 1997 to nearly 1.2 million metric tons in 2003—leaving Brazil ahead of Australia as the world's largest beef exporter by volume. Four-fifths of this growth came from the Amazon. During the 1990s, the number of cattle in Brazil nearly doubled. From 1990 to 2002, the Amazon's share of the country's total herd grew from about 18 percent to almost one-third—or 57 million head. The Amazon region was now losing 2.5 million hectares (9,500 square miles) of rainforest a year—almost half again the average rate of loss during the 1990s: 1.7 million hectares (6,500 square miles). The director-general of the Center for International Forestry Research, David Kaimowitz, seeing this statistical picture, concluded: "Brazil's deforestation rates are skyrocketing and beef production for export is to blame."[16]

Cattle ranching is directly responsible for over half of the deforestation in the Amazon region. But logging, clearing land for crops like soybeans, and small-scale subsistence farming remain core causes of deforestation as well. Logging in the tropics is commonly "selective," with loggers harvesting only the most valuable timber (old-growth trees). Although such harvesting rarely involves clear-cutting, it begins the process of deforestation by degrading forest integrity and biodiversity. Opening the canopy can dry forests out and, along with "kindling" littering the forest floor, leave logged forests more vulnerable to both natural and intentional fires (burning is a cheap and easy way to clear logged forests and in the process fertilize the soil). Logging roads can also provide ranchers, slash-and-burn farmers, and plantation companies with easier access to once remote land. Still, in the Amazon, where the total pastureland is almost 6 times larger than total cropland, cattle ranching is responsible for about 10 times more damage than logging.[17]

Despite continued cases of foot-and-mouth disease across the country (including in the Amazon region), beef exports remain strong. Brazil appears set to hold its ranking as the world's largest exporter of beef: it reached 40 new markets in 2004, selling $2.5 billion worth of beef to a total of 143 countries—$1 billion more than in 2003. Since then, beef exports have been regularly setting monthly records, both in revenues and in volumes, and are now worth about $3 billion a year.

A notable example of the capacity of Brazilian beef exporters to expand into new markets is Russia, which did not import much Brazilian beef before 2000. By 2005—just five years later—this market was worth over $500 million, accounting for over one-quarter of total Brazilian fresh and frozen beef exports.[18] Although beef exports earn valuable foreign exchange for Brazil's developing economy, their continuing growth is accelerating deforestation of the Brazilian Amazon. The total area of rainforest lost climbed in 2004 to more than 2.6 million hectares (10,000 square miles), before falling back in 2005 to just under 1.9 million hectares (7,300 square miles)—more in line with the rate of loss from 1999 to 2001—after commodity prices declined and news of an outbreak of foot-and-mouth disease spread.[19] Still, even with this slow-down, the deforestation rate in the Brazilian Amazon remains one of the world's highest.

The global cattle industry, then, is causing extensive ecological damage. Industrial fields of grain are covering an increasing share of the planet to feed these cattle. Antibiotics and hormones are seeping into local environments and through food chains. Manure is polluting local water-ways, and methane from the herds of bloated cattle in feedlots is pollut-ing the upper atmosphere. Ranchers are also carving tropical rainforests into vast "grasslands" and drawing down the world's oil reserves—a core cause of climate change—to subsidize the "efficiencies" of "produc-ing" ever more beef ever faster. This is not to diminish efforts to mitigate these ecological impacts. But, as chapter 17 will show, these efforts are trailing far behind the stampeding consumption of industrial beef.

17

Sustainable Beef? Chasing a Stampede of "Regular" Steers

Markets for natural, organic, and grass-fed beef are growing in many countries. Ranchers serving these markets tend to follow higher environmental standards for managing forests, land, water, and wildlife. They generally work on a smaller scale, treating cattle with more care, using fewer growth hormones and antibiotics, and avoiding both protein supplements from animals or fish and feeds from farms using pesticides, synthetic fertilizers, or genetically modified crops. The cattle tend to live in more sanitary and spacious settings and to eat a more vegetarian diet. And, finally, the beef generally contains fewer chemical preservatives, synthetic ingredients, and artificial colors or flavors.

Producing beef for such markets no doubt casts lighter ecological shadows. Yet the evidence in this chapter shows that the environmentally friendly methods many ranchers use to do so are not offsetting—or even slowing—the escalating global damage from consuming beef. One reason they aren't is the vague, confusing, and even misleading terms defining these markets. The label "natural beef" in the United States, for example, though sometimes a synonym for "organic beef" or "grass-fed beef," may, at other times, include beef from feedlot cattle treated with antibiotics and hormones. A second, far more significant reason is that consumption of natural, organic, and grass-fed beef still accounts for only a tiny fraction of the rising global consumption of beef, even where environmental markets are strengthening.

Markets for beef with environmental labels show every sign of growing and are even beginning to emerge in developing countries (sometimes with assistance from donors and nongovernmental organizations). Still, it seems unlikely these markets will appreciably increase their share of total beef consumption anytime soon. For one thing, many consumers are put off by the ambiguous labels, high price tags, and unexpected tastes of much organic beef. For another, government regulations and economies of scale favor big enterprises over niche producers. And, more important, multinational companies can make much bigger profits from selling large quantities of cheap, industrially produced beef to fast-food restaurants, supermarket chains, and the growing global middle class.

Defining All-Natural Beef

The meaning of "organic beef," as one Montana cattleman put it, is "slippery."[1] This is true, too, for the consumer labels "natural," "grass-fed," and "pasture-raised beef." The term "sustainable beef"—a label some producers are now using to describe a mix of organic, grass-fed, and industrial practices—is perhaps the slipperiest of all. Ambiguities arise for many reasons. These terms take on different meanings in different national and cultural settings. Jockeying among retailers for market shares can contribute to multiple meanings, too. And, finally, loopholes, exceptions, and lax enforcement can make it hard—even impossible—to really know whether a product actually meets the standards of *any* label. Thus, even in countries with strong regulations, these terms frequently confuse or mislead consumers.

Take the term "natural beef" in the United States. Companies like Coleman Natural Foods—which, after a merger in 2006, became one of the top 30 meat processors in the United States, with over 2,000 employees in 17 locations across 6 states—first began selling natural beef in 1979 as Coleman Natural Meats. The company's founder, Mel Coleman Sr., worked hard to build a market for his natural beef. Camping in a rented car, he drove from grocer to grocer to convince a few to stock his beef; he also supported a lobbying effort to get the Department of Agriculture (USDA) to establish labeling standards for natural beef, which it did in the mid-1980s.

The USDA, however, required only that "natural beef" undergo minimal processing and have no artificial ingredients, a standard well below Coleman's original understanding of natural beef. Today, the company asserts its "products adhere to a higher safety and ethical standard than products qualifying as 'natural' by USDA's definition." Coleman's all-natural beef is indeed "minimally processed." The cattle do not receive hormones, antibiotics, chemicals, or artificial ingredients. They live in "stress-reduced" and "spacious" settings. And they eat a vegetarian diet—"all-natural," although not organic. The company also imposes "strict environmental standards" for managing water, forests, land, and wildlife. This has been, according to the corporate message, the "Coleman definition of 'Natural' for more than three decades."[2]

Under USDA rules, "natural beef" demands far less adherence to principles of environmental stewardship. To qualify, it cannot contain artificial colors or flavors, chemical preservatives, or synthetic ingredients; and it must undergo only minimal processing—most notably, ground beef does not qualify. The Food Safety and Inspection Service also requires a brief definition of "natural" on the label. Processors and retailers define it differently, however, creating considerable confusion among consumers. Some, for example, label beef "natural" when the cattle are raised exclusively on a grassy range. Others define it to mean only that the cattle did not receive antibiotics or hormones. Still others use the term "natural" as a synonym for "organic." The former chairman of the USDA's National Organic Standards Board calls Mel Coleman Sr. "one of the pioneers to develop a truly natural meat company." Still, he laments, "the 'natural' label has lost its meaning" under current U.S. standards.[3]

Even though government standards for "natural beef" are fairly easy for industrial meatpacking firms to meet, until recently, most have stayed away from this niche market. This is beginning to change, however, as surveys show increasing consumer interest in healthy foods, and as these markets show signs of expanding. For example, three of the largest meatpacking firms—Tyson Foods, Swift, and National Beef—began to market "natural beef" in 2006.

Some small producers of natural beef are now worrying about their capacity to compete with these giants. "They process more cattle in one day," explained one natural beef producer in 2006, "than we do in one year." Already, these small producers struggle to get shelf space in supermarkets, and some are now planning to avoid competing with the big meatpacking firms and shift over to producing organic beef. Indeed, companies like Coleman's are already, according to Mel Coleman Jr., "one grain away from being organic."[4] Still, that one grain is not cheap: organic cattle feed can cost as much as 30 percent more than nonorganic. The result for American consumers is that all-natural beef is cheaper than organic beef—although both are more expensive than "regular" beef.

Defining "Organic" and "Grass-Fed" Beef

Under its 2002 National Organic Program, the USDA defines the term "organic" with more exacting standards than the term "natural."

Accredited agents inspect farms and certify producers, handlers, and processors as "organic." To qualify as 100 percent organic beef, the cattle must not receive growth hormones, antibiotics, or other prohibited medications. A sick animal treated with antibiotics loses its organic status (although vaccinations are fine). Producing organic beef cannot involve "most conventional" pesticides, synthetic fertilizers, or genetic modification. The feed must also be certified as organic—thus the cattle can eat grains like corn, for example, but this corn must come from organic farms.

The USDA's National Organic Program allows for three other uses of "organic" besides 100 percent organic. Retailers may label a product "organic" when it's made with at least 95 percent organic ingredients. They may use the phrase "made with organic ingredients" when the product contains at least 70 percent organic ingredients and no sulfites. And, finally, for a product made with less than 70 percent organic ingredients, retailers may use the term "organic" when listing ingredients. Only products that are "100 percent organic" or "organic" are allowed, however, to display the USDA "organic" seal. The rules for organic meat, along with higher prices, have kept the market small in the United States: the Organic Trade Association estimates that organic meat accounts for only 0.22 percent of total meat sales.[5]

Members of the American Grassfed Association have been trying for years to get the USDA to develop standards and procedures to certify grass-fed beef along the lines of the organic label. Many of these members see this as a higher (and better) standard than organic beef because grazing cattle on open pastures is more natural and appropriate than confined feedlots heaped with organic corn. In 2006, the USDA did propose a voluntary standard for the term "grass-fed," which would have required a diet of mother's milk and 99 percent grass, legumes, and forage. This would have still allowed ranchers to confine animals in feedlots so long as they fed them "grass"—defined as including hay, rice bran, and almond hulls. It would have allowed ranchers to use hormones and antibiotics, too. The reasoning, explained the chief of the standardization branch of the USDA, was to avoid diluting "the meaning of 'grass-fed'" and instead use separate standards for determining the extent of confinement as well as the use of hormones and antibiotics. The proposed standard for "grass-fed" also appeared to create a loophole that would allow farmers to classify "immature corn silage" as "forage."

The Grassfed Association angrily opposed this definition of "grass-fed." Because most people think of "grass-fed" as "cattle grazing on a pasture, not in a feedlot," many ranchers in the association saw such a standard as misleading. Most members oppose "grass feeding" in feedlots except in emergencies; and most also feel grass-fed cattle should remain free from hormones and antibiotics. Some members saw this USDA proposal more as a "logo" "for big companies" than a standard for small farmers trying to manage land sustainably. After first defending its definition and criteria, the USDA decided to withdraw its proposed voluntary standard for further consideration.[6]

Sustainable Beef?

Many ranchers find the standards for organic beef and for grass-fed beef as defined by the Grassfed Association hard to meet—or not worth the effort. Some of them are, instead, employing a mix of organic, grass-fed, and industrial practices, then labeling the product "sustainable beef." The Niman Ranch in California is a world leader here. Its motto is meat "Raised with Care." It works with over 500 independent "family farmers" who raise animals "traditionally, humanely and sustainably." The "goal" is "great taste." With a dash of hype, it claims to produce "the finest tasting meat in the world." The definition of "family" farm is narrow: "an individual or family owns all the animals, depends on the farm for livelihood and provides the majority of the daily labor necessary for the farm's management and upkeep." This ensures the participating farms are small enough to allow for sustainable practices. These farms should not, for example, need to liquefy or store waste in cesspools, a common source of noxious gases and water pollution from industrial farms. To qualify as "sustainable," these small-scale farmers or ranchers must practice "environmental stewardship," meeting Niman's "strict" environmental, health, and breeding "protocols," including passing inspections.[7]

Currently, the U.S. government has not set standards for sustainable farming, but ranchers, distributors, retailers, and chefs elsewhere in the United States are apparently following similar principles of management. The aim of some in this network—in the words of one chef serving "sustainable beef"—is to leave the "smallest and lightest footprint" possible.[8]

Yet what defines "sustainable" on these family farms is even more ambiguous than the terms "organic" and "grass-fed." Some common characteristics do exist. The cattle tend to spend less time in feedlots: ranchers breed them later in the season than traditional farms to ensure plentiful spring grass for the calves. Ranchers tend as well to allow these calves to roam more freely to reduce both stress (improving the quality of the beef) and the chance of contracting diseases. They do not use growth hormones, and the cattle tend to live about six months longer than on industrial farms (large packinghouses tend to slaughter cattle at 12 to 14 months). A small number of these ranchers market the beef as "free-range and grass-fed." Some, too, use organic feeds or maintain organic pastures. But most, after 14 to 16 months of grass feeding, fatten their cattle for 4 to 5 months on nonorganic grain diets in feedlots. Ranches like Niman do not produce government-certified organic beef, primarily because, according to its owners, feeding cattle organic grain (in short supply in the United States) would increase the cost of production by as much as 50 percent.[9]

In the case of Niman Ranch, however, feedlot cattle have access to shade and sprinklers and more room to move around than cattle in industrial feedlots. The workers also make efforts to ensure what the ranch calls a "humane end" for the animals—in part, because adrenaline from anxiety and stress during slaughtering can make their meat tough. In addition, there's a commitment to "vegetarian" feeds even though the USDA allows some animal by-products in cattle feeds like chicken feathers and fish meal (bone meal was banned in 1997). Moreover, unlike organic farmers, sustainable ranchers do not forgo all fertilizers and pesticides. Some might choose, for example, to spray a pesticide to kill weeds in ditches rather than take the time to pull them by hand. When pressed, some might also feed cattle hay grown with chemicals or fertilizers. Sustainable ranchers will treat sick cattle with antibiotics, but will not put antibiotics into the feed or water as a preventive measure or to stimulate weight gain.

Raising "sustainable beef," ranches like Niman stress, allows ranchers a way of leaving the smallest and lightest footprint that is simpler and more practical than raising organic beef—by avoiding pesticides, fertilizers, and antibiotics *whenever feasible.* Niman Ranch also labels its beef "natural" even though it admits the term has little real meaning for consumers: the USDA's focus on the degree of processing and use of preservatives is at odds with common sense, allowing, for example, meat from animals fed antibiotic feed to receive this label. Niman Ranch stresses that, when applied to its beef, however, "natural" means something different—a promise to use "all-natural" feed and a commitment to allow animals to experience "natural behaviors."[10]

Chasing Beef Markets

Over the last decade, consumer demand for sustainable, grass-fed, and organic beef has been growing in wealthy countries. The revenues from producing sustainable meat at Niman Ranch, for example, jumped tenfold from 1997 to 2003. Over 1,000 farms in the United States now specialize in raising grass-fed cattle in pastures—a twentyfold increase from 2001. In 2005 alone, natural and organic beef sales grew in the United States by 17.2 percent—to almost $120 million—compared with an average growth of 3.3 percent for total beef sales. That same year, retail organic beef sales earned $49 million, up from $10 million in 2003. Still, all of this represents fewer than 1 percent of the 33 million cattle consumed in the United States each year.[11]

The organic beef market in Europe would seem like one with even more potential to expand. The European Union provides a common definition of the term "organic," requiring a license from a government-approved certifier to produce or sell products with this label.[12] Countries like Germany are encouraging the use of organic labels to help reorient agriculture away from industrial "agrofactories" and toward organic farming.[13] The middle classes of Europe are also well educated and wealthy by world standards. Governments in Europe—compared to, say, those in the United States—tend as well to proactively manage environmental and health risks from food (by banning hormones in beef and genetically modified organisms in cereals, for example).[14] Moreover, mad cow disease, which began in Britain in the 1980s, has been undermining consumer trust in "regular" beef across Europe for decades now. One Danish butcher, following a crash in business after the surfacing of mad cow disease in France, Germany, Spain, and Italy in 2001, lamented: "No one even asks for red meat anymore. Shoppers are terrorized."

Many of the consumers worried about mad cow disease were not, however, necessarily forgoing meat, but were replacing beef with pork, chicken, lamb, and fish. Sales in exotic meats—such as ostrich, emu, kangaroo, and bison—began to rise, too. Some consumers were turning to organic beef to avoid eating any cattle fed bone, brain, or spinal scraps of other cattle (widely reported as a source of mad cow disease). "The mad cow crisis," explained one Parisian greengrocer experiencing rising sales in early 2001, "has been a real shot in the arm for organic producers."[15]

The United Kingdom was far and away hit hardest by the mad cow disease, with more than 175,000 cases. Over 80 people died of the terrifying human variant, Creutzfeldt-Jacob disease, where the brain wastes away. The United Kingdom would therefore seem like a particularly promising market for organic beef. Yet this has not been the case. The mad cow panic sweeping the European Union in 2001, which caused beef consumption across the region to plunge 27 percent in the first month of 2001, did little to rattle UK consumers. Beef purchases actually went up by 3 percent in January 2001.[16] Organic beef sales have gone up modestly since then: for example, about 2,250 organic cattle were sold in 2003, up 250 from 2002. But, as in the United States, this remains a tiny portion of the meat market, in part because, according to the managing director of the UK Organic Livestock Marketing Co-op, many consumers are not willing to pay the current "premium for organic meat."[17] That premium is often steep, with many organic cuts in the United Kingdom costing as much as twice their nonorganic equivalents. Organic beef sales in the United Kingdom are not even keeping pace with the more robust increase in sales of organic foods more generally—and of imported organic fruits and vegetables in particular.[18]

Consumers in the United Kingdom seem unwilling to pay a premium for organic beef for many reasons. The quality of industrial beef is now higher than during the mad cow days; and "farm assurance" programs have reassured consumers that industrial beef is now safe. Some consumers, too, are wary of paying up to twice as much for beef just because of a "label" or a butcher's promise. Documentaries like the 2006 West Eye View series exposing butchers who are selling "regular" beef as "organic" have further undermined public trust in such claims. Moreover, because the government doesn't allow farmers to use growth hormones, many consumers in the United Kingdom continue to perceive industrial UK beef as "green." Some would go so far as to argue that it's "better to buy non-organic from the United Kingdom rather than something organic from round the world."[19]

Other factors besides consumer preferences or a willingness to pay premiums are impeding organic beef sales, too. Supermarkets are the main outlets for organic beef sales in the United Kingdom: about 85 percent of organic beef and lamb sales go through large retailers. Many supermarkets are only willing to sell top cuts (such as sirloin steaks or roasts) with guarantees of consistent supplies (some require suppliers to promise future deliveries). Many are unwilling to carry organic burgers—in part, because supplies are too small for the turnover necessary for profits. This barrier is so arduous the UK Organic Livestock Marketing Co-op estimates that 60 percent of organic forequarter meat for sausages, mincing, or stewing is still sold into the industrial meat market without an organic label.[20]

Similar barriers to expanding sales of organic and pasture-fed beef exist in other markets as well. Regulators everywhere tend to design standards that are easier for big companies to meet. It's harder, too, for small organic ranchers to find buyers. Large slaughterhouses and processing facilities prefer to work with large numbers of cattle. Small ranchers, whose meat-processing facilities are also small, struggle to compete against industrial farms processing thousands of cattle a week.

Small ranchers in the United States find it difficult to meet the government standards for "organic" beef because less than 2 percent of U.S. grain is organic. The cattle bred for the feedlots, moreover, do not always fare well on grass, and bad weather can disrupt feeding plans. The grass also needs a chance to grow back, and, without careful management, the quality of both land and cattle can deteriorate over time. Much of this grass-fed beef is also sold frozen because of the focus on local sales in small quantities; much of it is from cattle slaughtered in the autumn before less optimal grass conditions slow the rate of weight gain. The result can be beef that North American consumers find neither "tender," "juicy," nor "tasty," used, as so many of them are, to corn-fed beef.[21] "It doesn't even taste like beef," a Vancouver chef remarked in a 2005 interview. "It tastes like eating grass."[22]

Still, despite all of these hurdles, organic ranching is beginning to emerge even in developing countries, in part, because of various efforts to assist small ranchers. The organization Heifer International, for example, is helping communities build processing facilities and working to convince consumers and farmers of the benefits of producing and eating local beef.[23] The World Bank, too, has put in place policies to support small farmers and organic food in developing countries. In a program funded by the nonprofit organization Conservation International and Brazil's Biodynamic Beef Institute, six cattle ranches— covering 160,000 hectares (about 400,000 acres)—now raise certified organic beef on native grasses in the Pantanal region of Brazil's Mato Grosso. To qualify, the ranchers must raise native breeds without antibiotics or growth hormones and must prevent cattle from overgrazing or destroying local vegetation.[24]

Grazing on native grasses can reduce the ecological and energy reserve costs both of growing corn and soybeans with chemical fertilizers and of transporting grain to feedlots. But, as the analysis of Brazil in chapter

16 shows, "grass ranches" can also cause deforestation. Moreover, over-grazing can cause soil erosion, loss of productive agricultural land, and eventually desertification. And even though, on balance, raising beef in environmentally friendly ways *does* lighten the shadow effects of con-suming beef, as the analysis in this chapter shows, the growing markets for sustainable, organic, and grass-fed beef still serve only a tiny propor-tion of consumers—a trend unlikely to change anytime soon.

18

The Globalization of More Meat

Producers and retailers respond to consumers—to their desires, tastes, rising incomes, and, most concretely, to their actual purchases—because more customers means more profits. Yet, as the case of beef shows, the notion that demand arises from the innate needs and cravings of freethinking consumers—even for a basic need like food—is far too simplified.

Governments can shape consumer demand by declaring certain things illegal, by limiting choice with restrictive trade or social policies, by imposing differential taxes, and by setting labeling standards. They can do so through incentives, as well, by subsidizing producers to develop or expand markets, causing surpluses of grain, for example, which in turn makes grain feed a cheap way to fatten animals. Over time, this begins to define consumer tastes, turning the marbling and texture of corn-fed beef into what consumers expect, prefer, and demand in their beef. Management systems and technologies like soybean plantations, feedlots, and disassembly lines can fill markets with increasing amounts of cheap marbled beef, making supersize meals profitable for a globalizing fast-food industry. Fast-food advertisers can then market "big" and "more" as a deal, as irresistibly satisfying to the increasingly obese consumers of the world.

The consumption of meat in a globalizing world has been rising steadily for over a century. Per capita consumption of meat was rarely above 5–10 kilograms (about 10–20 pounds) a year in most traditional agricultural societies. By 1950, it had reached 17 kilograms (38 pounds) a year worldwide. Today, it's over 40 kilograms (88 pounds) and rising, with some of the wealthiest countries having annual per capita rates of more than 125 kilograms (275 pounds). Total meat consumption is now over 260 million metric tons and increasing steadily, as people eat more and the world's population grows by more than 200,000 every day.[1]

Consuming so much meat is casting ecological shadows over rural ecosystems, global water and food supplies, tropical rainforests, and the earth's climate. Billions of animals are multiplying their numbers on industrial farms. To produce more meat more efficiently, feedlots are flooding local ecosystems with antibiotics, hormones, and animal waste. Plantations for animal feed like corn and soybeans are relying on genetically modified seeds as well as on chemical pesticides and fertilizers to ensure cheap crop surpluses. With the technical and financial assistance of multinational corporations, plantations and ranches in places like the Brazilian Amazon are clearing rainforests—hotspots of biodiversity—to increase exports of cattle and soybeans for beef consumers worldwide, from Canada to Chile to Europe to Egypt to China.

Of course, not all beef producers and governments are ignoring the ecological consequences of the cattle industry. Some governments now have eco-labeling programs for beef, and some ranchers are taking independent steps to produce beef more sustainably. Consumers in countries like the United States can now choose from among all-natural, organic, sustainable, grass-fed, and pasture-raised beef.

Eco-labeling programs are shifting ranching practices and consumer preferences in North America and Europe. Efforts are also under way to expand the capacity of developing countries to produce organic beef, by, for example, providing financial assistance to organic farmers. Together, these developments would seem to represent an encouraging market trend, yet markets for sustainable, organic, and grass-fed beef still account for less than 1 percent of total beef consumption, even in the United States, where market growth for such beef is "strong." Overall, these markets are doing little to diminish the ecological shadows cast by rising demand for beef from industrial ranches.

Many factors are limiting the growth of these markets. Definitions of terms on eco-labels vary from product to product, confusing some consumers. More exacting standards for eco-beef have pushed its prices well above those of regular beef. With such a small share of the market, and working with relatively small herds of cattle, eco-beef producers struggle to compete with industrial farms and multinational meatpacking firms able to process far more, faster and cheaper. The large slaughterhouses don't see any point in handling the small orders of eco-ranchers, and the large supermarkets don't offer much shelf space for high-priced eco-beef. The average consumer is reluctant to pay twice as much for a steak with an eco-label, especially when the meaning of "organic" or "natural" or "sustainable" is ambiguous. Finally, governments are offering only token support for these niche markets, focusing instead on stimulating more consumption of industrial beef to maintain national economic growth and a healthy global economy.

In the first four cases of this book—the automobile, gasoline, the refrigerator, and beef—consumption has continued to rise over many decades as firms have modified their practices in response to regulatory reforms and market shifts. For some consumer goods, however—whale meat, elephant tusks, and seal furs, for example—it has been impossible to keep consumption rising as producers hit ecosystem limits. As the following history of the Canadian seal hunt shows, the resulting declines in economic and political power can allow counterforces—in this case, animal rights and environmental activists—far more influence over the direction and intensity of ecological shadows than for products like the automobile or beef. But, as the case of sealing also shows, such influence can quickly dissipate when commercial interests are able to enter markets beyond the reach of these activists.

V

The Harp Seal Hunt

19

To the Red Ice: Heroes and Overharvesting

The seal hunters of the eighteenth and nineteenth centuries in what is now Atlantic Canada were heroes of local folklore by the beginning of the twentieth century, men who endured the hardships of the hunt to eke out a living for their families. From the 1960s into the 1980s, the global outcry against the inhumanity of the seal hunt turned this history inside out. The consumers of luxury furs increasingly began to see these onetime heroes—called "swilers" by Newfoundlanders—not as hunters on a hunt, but as barbaric men on a rampage, clubbing bawling baby seals to death, stealing the beauty from a white spring morning.

To the shock and anger of the swilers, the whole world seemed to turn against them. Conservationists were declaring the harp seal population to be endangered by overharvesting. Government officials were imposing one regulation after another. Activists were interfering with the hunt. And markets were becoming unstable as increasing numbers of consumers began to look upon the life of a swiler with disgust. Commercial sealing in Newfoundland went into decline. All of this made little sense to the swilers, who saw the hunt for "whitecoat" harp pups (from 6 to 12 days of age) as more, not less, humane than cattle, pig, or sheep farming.

Then, in 1983, the European Community—by far the largest market—imposed a two-year ban on the import of whitecoat pelts. It was the beginning of the end. The European ban was renewed in 1985. With an anti-sealing boycott of Canadian fish products gaining momentum in both Europe and the United States, the Canadian government surrendered to the global outcry in 1987 and banned the centuries-old spring hunt "to the ice" for whitecoats.

To understand why this happened, we need to compare the emerging power of activist groups from the 1960s to the 1980s, with the relative political and economic decline of sealers during this period. Doing so reveals complex, often subtle and hidden, ways that the globalization of environmentalism and the globalization of consumer markets can interact to shift ecological shadows of consumption. Examining why the commercial hunt of older seals resumed in the mid-1990s adds further to our understanding here.

Setting the stage for this analysis, this chapter sketches the history of the commercial seal hunt from its beginnings in the eighteenth century, through its heyday in the nineteenth century, to the start of the activist anti-sealing campaign in the 1960s. It focuses on the impact of changes in hunting practices, processing technologies, and consumer markets on the sustainability of the harp seal herds. And it shows how, as entrepreneurial traders became increasingly efficient at bringing home this natural resource, some ecological shadows of consumption were forming and lengthening centuries ago.

It begins with the ill-fated voyage of the SS *Newfoundland* in 1914, a voyage honored in the songs and legends of Newfoundland as a tale of courage and hardship—one of the many chapters in the story of sealing that animal rights and environmental activists would later rewrite as one of brutality and depravity.

The Story of Albert John Crewe

When the wooden SS *Newfoundland* steamed out of St. John's harbor on a black midnight in early March 1914, Albert John Crewe was just 16 years old. Along with his father, Reuben, and hundreds of others, Albert was heading to the ice off southeastern Labrador and northeastern Newfoundland—what the swilers called "the Front"—to hunt seals.

The older swilers, like 49-year-old Reuben, knew what dangers lay ahead. On the 1911 hunt in the Gulf of St. Lawrence, ice floes had crushed Reuben's steamship, the *Harlaw*, during a storm. Somehow, he'd managed to scramble over the ice and up the sharp cliffs of St. Paul's Island. When he finally made it back alive, he vowed to his wife, Mary, never to return to the ice again.

But then, just three years later, his oldest son, Albert, was offered a berth on the *Newfoundland* under Captain Westbury Kean. Albert was thrilled: Captain Wes Kean was a rising star in the sealing fleet and the son of Abram Kean, the most famous sealing captain of all time (who, in 1934, after 45 years of swiling, would land his millionth seal). But Reuben didn't want Albert to ever join the seal hunt: swiling on the jagged and shifting "pans"—floating sheets of coastal ice—was just about the toughest work imaginable; the sealing industry had a long history of exploiting men eager to scratch anything at all out of the harsh landscape of their birth.

The hunters, Reuben knew, would be seeking primarily "whitecoat" harp pups from 6 to 12 days of age, whose pelts (skins with the fat) were the most profitable.[1] To reach these seals, the wooden *Newfoundland* would need to navigate treacherous ice pans, which collided and ground together with the power to crush the ship as though it were made of toothpicks.

Weather permitting, the captain would search for patches of white-coats and, if lucky, sight the "main patch," which he'd announce by hollering "Swiles!"[2] He would then wedge the boat against the ice and send the swilers over the side to kill the whitecoats. The trek across the jagged and shifting ice pans would test Albert's strength and agility. He would need to learn how to clamber over peaks and ridges of ice with his gaff and towrope and how to leap from pan to pan—a skill known as "copying."

The dangers would be constant: a fall into the freezing Atlantic could mean death in a matter of minutes. Sudden snowstorms and fog could strand sealers on the ice at any time. Once he reached the whitecoats—once he was what the swilers called "in the fat"—Albert would need to overcome any instinctive sympathy for the bawling week-old fluffy white pups lying helpless on the ice. Standing over one, he'd need to raise his club high above his head. And even though the pup's large brown eyes would stare at him and seem to pour tears, he'd need to strike with enough force to crush its skull in a blow or two.

Afterward, with the pup's swimming reflexes still quivering, he'd need to take his sculping knife and slice the pelt from the carcass as fast as possible. There'd be no time to feel nauseous, much less worry whether the pup was really dead—time was money for the swilers, the captain, and the shipowners. A "good hand" was expected to haul in 120 pelts a day; a few, in ideal conditions, were said to take 300.

Pride was at stake, too. The first vessel to reach port with a full load of pelts would win the title of "highliner"—a glory some captains and crews would take great risks to achieve, such as running a ship loaded with heavy pelts through bad weather. Newfoundlanders would be betting on which vessel would win. The captain of the highliner ship would receive a silk pennant (a tradition that began in 1832); the crew might win prizes, too, perhaps even a crate of oranges.[3]

Albert would have little time to rest once the killing began. The swilers would need to haul the heavy pelts with towlines either to collecting pans or back to the ship. Then the men would need to prepare the tens of thousands of pelts for storage for the trip home. These would soon bathe the ship and all the men in grease. When the supplies of salted cod and pork ran low, the men would eat seal hearts, seal livers, and boiled flippers. The stench of seal would seep into every cranny of the boat; the cramped sleeping quarters would become filthy, slippery with fat and blood. Once the hold was full, the captain might even force the crew to sleep on top of greasy pelts.

The trip would indeed be fraught with infernal dangers and hardships—no place for a lad. The night Albert burst in with the news that his Uncle Ben had secured him a berth of the *Newfoundland*, Reuben and Mary tried hard to dissuade their son. But Albert fought even harder to go. He saw the trip as a grand adventure, a way to prove his manhood, to join the generations of Newfoundlanders brave enough to go to the ice. By evening's end, his parents relented: Albert could go to the ice, but Reuben would join him and remain at his son's side the entire time.[4]

Adventure on the Ice

Albert's grand adventure started inauspiciously. When the *Newfoundland* became locked into the ice for days on end, the crew suspected fate might be against them. All the more so after two stowaways—a surefire way to anger the spirits of bad luck—were found on board.

The men labored with steel-tipped poles to break up the rafted ice piling against the sides of their ship. They shoved dynamite under the ice, whose southward drift kept pushing them backward. Then, through the now exploded and buckling ice, the *Newfoundland* inched its way toward the larger steel steamers of the main sealing fleet. Finally, early in the morning on the last day of March, frustrated at not making greater headway, Captain Wes Kean sent 166 of his men walking across the ice in search of patches of whitecoats many miles off.

The sky, to many of the experienced swilers, began to look threatening. Mid-morning, after trekking four miles over rough ice, 34 of the men, convinced a storm was brewing, turned back to the *Newfoundland*. This they did to shouts of "Cowards!" from the others, who pushed onward, some with bravado, others simply obeying orders and trusting their team leaders and captain, toward the *Stephano*, a steel steamer under the command of Abram Kean.

No one will ever know what was going through the minds of Reuben and Albert Crewe. But both went forward. At 11:20 a.m., their group managed to reach the *Stephano*. Then, while feeding the men a quick meal, Captain Kean steamed away from the *Newfoundland* toward a patch of whitecoats. He dropped the men over the side, some still exhausted from their trek to the *Stephano*, instructing them to sculp the whitecoats and then hike back to the *Newfoundland*.

Within hours, a blizzard was raging. Reuben and Albert, along with 130 other men, fought their way across jagged and shifting ice pans in search of the *Newfoundland*. Captain Kean had dropped them far from the *Newfoundland*—much farther than anyone thought, including Kean himself. Lost and exhausted as night fell, the men built a wall of ice and snow, huddling behind it as snow and rain and sleet whipped around them.

Somehow, Reuben and Albert managed to survive a night when men froze into lumps on the ice pan all around them. By mid-morning, however, both could go no farther. Albert lay down. His father lay next to him, holding on tight, tucking Albert's head under his thick Guernsey sweater in a desperate bid to keep his son warm. They died a short while later, frozen together.

After two days and two nights in this hell, two men, their limbs dead and swollen, staggered to the steel steamer *Bellaventure* (no one was searching for the men because captains Wes and Abram Kean each thought the men were on board the other's ship). Alerted, search teams hurried across the rough ice to find survivors in excruciating pain, some stumbling about in a daze, others huddling behind stacks of frozen bodies for shelter. A silent and solemn crowd of 10,000 waited all day in St. John's for the return of the *Bellaventure*. Reuben and Albert were only two of the 78 men to die in the "Great Newfoundland Sealing Disaster of 1914."[5]

Wooden Tragedies

This tragedy is just one in a long series for the Newfoundland sealing industry. The same storm that took Reuben and Albert sank the *Southern Cross* with all hands (173 men) as it raced from a hunt in the Gulf of St. Lawrence, heavily loaded with whitecoat pelts, for the honor of being highliner of the fleet.[6] Weather was only one of the many hazards of sealing. The ice could, as Reuben Crewe tried so hard to tell his son, trap and crush the wooden sealing ships with little warning. Boilers could explode at any time; so could stores of dynamite and gunpowder. It was routine to "lose" ships—a cost of doing business. Over the course of the nineteenth century, more than 400 wooden sailing vessels were lost during the seal hunt. One of the worst disasters of this era was a hurricane in 1832 that destroyed 14 schooners and killed over 300 men. The number of wooden steamships—which locals called "wooden walls"— lost in the latter half of the nineteenth century reveals the terrific dangers of sailing in these ships and the owners' near-total disregard for the safety of swilers. The first wooden steamer went to hunt seals in 1863; by 1900, 41 of the first 50 to go sealing were lost at sea. The early twentieth century was no safer: between 1907 and 1912, 11 more wooden walls went down during the spring seal hunts.[7]

At least 1,000 Newfoundlanders have died from unsafe ships, the treacherous ice, and the freezing Atlantic during the brutal history of this industry. Understanding how it came to sacrifice men for pelts requires us to step back and survey the biology, technology, and consumption shaping the sealing industry since the first commercial hunters of the eighteenth century.

The Biology of Sealing

Every year, harp seals migrate in loose herds of a few hundred between arctic and subarctic regions of the Atlantic. They spend the summers in the eastern Canadian Arctic and off western Greenland. The northwestern population then travels from the Davis Strait to whelping grounds off southeastern Labrador and northeastern Newfoundland (the Front) and in the Gulf of St. Lawrence (known colloquially as "the Gulf") east of the Magdalen Islands.

Females generally give birth to a single pup on the ice floes in late winter and early spring. The pups are thin, yellow, and scraggly at birth. The rich milk of mother seals allows the pups to fatten at a rate of over two kilograms (around 5 pounds) per day. After six days or so, the pups have soft, thick white coats that last until about 12 days of age. These whitecoat seals are placid and sleep for long stretches in the sun, generally avoiding the freezing water until they have enough blubber to survive. The mother seals, who generally do not defend their pups, wean them after 12–15 days, then head off to join the adult males for mating. After weaning, pups lose fat and thus potential oil.

The pups begin to shed their long white fur between 10 and 14 days of age—rubbing against the ice as if to scratch a nasty itch. The coat becomes ragged and coarser until they finish molting 2–3 weeks later. Sealers call the pups during the molting stage "ragged jackets." Once a pup fully molts, it's called a "beater" until it's one year old. Beaters have sleek silvery coats with black spots.

The Beginnings of a Sealing Fleet

The First Nations peoples of what is now known as Newfoundland and Labrador relied on harp seals for subsistence for thousands of years before the first Europeans began to hunt them in the sixteenth century. Native peoples and early European settlers alike ate seal meat and burned seal oil for light and warmth. Sealskins were made into warm winter hats, jackets, mittens, and boots. The first commercial sealing began when the British set up a sealing post on the Labrador coast in 1765. This post and posts in Twillingate and Bonavista Bay on the northeast coast of Newfoundland generated an average income from seal oil exports of nearly £10,000 per year in the 1770s.

Seals seemed as plentiful as blackflies to the early explorers. A French sealer in 1760, coming upon a herd of seals sweeping over the horizon, wrote that it took 10 days to sail past. Until the 1790s, most seals were caught by "landsmen" using nets close to shore in winter, when southward migration during the winter months brought the seals near land.[8] The seal catch was small over the eighteenth century: with an annual average of about 27,000 from 1723 to 1803. By the end of the 1700s, however, commercial sealing was starting to take off as schooners and brigs, with crews of up to 50 men, began to sail to the ice in spring.[9]

Sealing in the Nineteenth Century

The nineteenth century was the heyday of the Atlantic sealing industry. Sealing accounted for up to one-third of Newfoundland's total exports in the first half of the century; only salted cod exceeded seal oil in terms of export value for Newfoundland over this century.[10] Sealers managed to land over 500,000 pelts in 11 of the 35 years from 1825 to 1860. Sealers off Newfoundland landed, on average, 470,000 seal pelts per year from 1831 to 1840 and 440,000 pelts from 1841 to 1850. The all-time high was in 1832, when sealers took over 740,000 seals during good weather conditions.[11] The biggest hunting effort ever was in 1857, when 370 vessels and 13,600 men went to the ice off the northeast coast, although sealers were able to land "only" 500,000 seals.

The main seal products at this time were seal oil and sealskins in that order; the primary market was the United Kingdom. Seal oil was used for soaps, lamps, and lubricants (such as for sewing machines). The thin seal hides often ended up as fine leather for wallets, handbags, bicycle seats, shoes, boots, cigar cases, and bookbinding. There was no market for whitecoat furs yet because sealers didn't know how to treat the lanugo fur to prevent it from falling out, but demand for the oil and fine leather from the 6- to 12-day-old whitecoat pups was high.

Hunters during this time relied mainly on a 3-foot wooden club or a gaff—a wooden pole with an iron hook—to kill the whitecoat seals. The gaff was the more common tool because it helped sealers test the ice, keep themselves steady when leaping from pan to pan, hook onto the ice should they fall into the sea, and haul other men out of the water and onto boats.

The ability to catch whitecoats depended on the weather and the condition of the ice. Sealers would leap from pan to pan when the ice was broken up—or sometimes use punts to transport men and pelts across open patches of water. They did their best to fill the ship's hold with whitecoat pelts, but this was often impossible, so they would turn to shooting beaters and adults with long sealing guns from a punt. Sometimes, after filling a hold with whitecoats and dropping these off at a home port, sealers would return to the ice later in the season to hunt beaters and adults.[12]

Under the onslaught of such large commercial hunts, the population of the northwestern Atlantic harp seals began to decline in the second half of the 1800s. The average annual seal catch off Newfoundland from 1851 to 1860 fell slightly, to 430,000, then dropped to 290,000 from 1861 to 1870.[13] In the last third of the nineteenth century, environmental pressures intensified as wooden steamships, more maneuverable than schooners and with crews that could exceed 200, joined the Newfoundland seal hunt. The number of wooden steamers rose steadily, from 18 in 1873 to 27 in 1880–81. (The first steel steamers, with hulls shaped to ride onto the ice and crush it, would not join the seal hunt until 1906.)

Wooden steamers were more efficient than sailing vessels, requiring half the crew to obtain the same number of seals. These larger and more productive ships also allowed larger firms in St. John's to take control of the sealing industry by the end of the nineteenth century.[14] Because the herds were now smaller, however, seal catches didn't reach the heights of the 1830s and 1840s. The average annual catch off

Newfoundland between 1881 and 1900 was 270,000.[15] By the end of the century, a total catch of over 300,000 was a good year—hundreds of thousands less than an average year in the mid-1800s.

The Decline of Sealing: 1900–1945

Hydrogenation, which saturates unsaturated (liquid) oils with hydrogen, turning them into saturated (solid or semisolid) fats, created new markets for seal oil, such as for margarine and chocolates, in the first half of the twentieth century; nevertheless, the sealing industry continued to decline. The average harp seal catch from 1895 to 1911 was just 249,000.[16] World War I reduced the Newfoundland sealing fleet from 20 vessels to just 12 in 1918 (all wooden steamships). By 1923, only two firms— Bowring Brothers and Job Brothers—continued to hunt seals. The Great Depression further eroded the Newfoundland sealing industry, and only four vessels, with fewer than 100 men, went to the ice in 1932. From 1915 to 1936, the Newfoundland sealing fleet managed to surpass 200,000 pelts in only 6 of 11 years; from 1912 to 1940, the average annual catch of harp seals fell to just 159,000.[17]

For several reasons, the number of landed seal pelts was less— sometimes far less—than the number of seals killed. Many beaters and adults were "struck and lost," swimming away mortally wounded or sinking before sealers could gaff them out. Typically, sealers would creep up on the more mature seals and shoot them on the ice, but, even then, some of those shot in the lungs or neck would tumble into the ocean. Captain Abram Kean estimated that as many as twenty adult seals were "lost" for every one caught during the sailing era.[18]

The practice of "panning," which began during the wooden steamer era, added to the loss of many whitecoat pelts as well. Instead of towing pelts back to the ship, sealers would gather them on large and stable "collecting" pans, marked with a flag bearing the ship's insignia. This allowed them to catch more whitecoats during daylight hours. Sometimes sealers would mark the pan with a kerosene torch so ships could collect the pelts at night. Panning proved to be extremely wasteful, however: changing weather or ice conditions could make it impossible even for steamers to collect pelts from the pans; thousands of pelts could simply disappear if the ice shifted and waterways opened up.

By the 1920s, some people were calling on shipowners to halt the hunt for a few years to allow the northeast Atlantic seal herd to recover.

Already, as far back as 1887, the Newfoundland government had imposed a 12 March–20 April hunting season for steamers and prohibited second trips after 1 April. The intent was both to allow sailing vessels to compete with the faster steamers and to ease the commercial pressure on adult breeder seals. In 1892, the government shortened the season for steamers by two days and banned *all* second trips. In 1916, in another effort to ease the death toll among breeder seals and to keep the focus on whitecoats and beaters, the Newfoundland government restricted the number of rifles to two per ship. In 1931, the government banned rifles altogether, although, later, ships were allowed one rifle apiece.

Such actions, along with World War I and the Great Depression, may explain why the population of harp seals appeared to stabilize in the 1930s. World War II again took pressure off the seal population when most fit men went overseas and sealing steamers were pressed into military service. Fewer than 1,000 Newfoundlanders, for example, took part in the 1941 hunt (the lowest number since 1932). This respite in the commercial hunt at the Front and in the Gulf allowed the population of harp seals to recover somewhat by the end of World War II.

Just before World War II, the United Kingdom was importing 72 percent of Newfoundland sealskins and the United States 26 percent. By 1946, the United States was importing 61 percent of Newfoundland sealskins, Canada 21 percent, and the United Kingdom 18 percent. The market for seal oil—by then, an ingredient in a wide range of products from chocolate, margarine, and nondairy whipping cream to cosmetics and machinery lubricants—shifted even more than the one for sealskins. To assist the Canadian war effort, Newfoundland diverted its seal oil to Canada (which it would not join until 1949); by 1946, it was exporting 98.5 percent of its seal oil there.[19]

Sealing toward a Crisis: 1945–1965

Because the Norwegians had found ways to prevent the lanugo fur from falling out of a dressed fur, the market for whitecoat furs began to expand after the war, although this didn't halt the decline of the sealing industry in Newfoundland during the 1950s. The decline reflected broader social trends that arose from Newfoundland's joining Canada in 1949: sealing was no longer as vital to the new province's economic welfare. But Norwegians and Nova Scotians soon began to replace Newfoundlanders at the seal hunt.

The first Norwegian vessel to join the hunt went to the Gulf of St. Lawrence in 1913. But interest soon waned, and after World War I, the Norwegians went back to their traditional hunting grounds in the White Sea near the Soviet Union. By the late 1930s, however, after the stocks of seals in the White Sea and off Jan Mayen Island near Greenland had declined, Norwegian sealers were arriving in force at the Front.[20] By the 1960s, more than half of the ships at the large-vessel hunt were from Norway. The Nova Scotian sealing industry was flourishing, too; by 1954, more ships went to the ice floes from Nova Scotia than from Newfoundland. In 1961, only 4 of the 28 vessels at the Gulf and the Front were from Newfoundland (although Newfoundlanders were part of the crews of non-Newfoundland vessels).

The Norwegians and Nova Scotians brought with them better ships and better equipment, such as refrigeration, radar, and helicopters. As the hunts became increasingly efficient, the average annual catch went up, to 255,000 seals (excluding the average landsmen harvest of about 55,000) in 1946–61, or almost two-thirds more than the average annual catch of the sealing fleets during the steamer period (1929–39): 156,000.

To maintain such large catches, sealers in the post–World War II era were hunting more mature seals. Not bound by Canadian regulations, the Norwegians continued to hunt after the official end of the Canadian season, a date designed to protect adult breeder seals during their northern migration in May (and thus ensure a healthy whitecoat population for the next season). Before World War II, whitecoats had accounted for about 90 percent of the harp seal catch of the Newfoundland fleet; by 1955, they accounted for only 60 percent of the entire northwest Atlantic harp seal catch.

The postwar resurgence of the Atlantic sealing industry led to a further decline in the population of northwest Atlantic harp seals in the 1950s. An aerial survey in 1950–51 put the population at 3.3 million (with 645,000 pups—215,000 in the Gulf and 430,000 at the Front). Little was done to conserve the stocks, however, and the average annual harvest from 1951 to 1955 was close to 330,000 seals. As stocks fell, the harvest tailed off to just over 300,000 per year from 1956 to 1960. Canadian sealers pointed to the increasing number of breeder seals killed—twice as many as before the war— blaming Norwegian vessels for hunting seal herds well into May. The next survey in 1959–60 put the total Atlantic harp population at just 1.25 million (with only 360,000 pups—150,000 in the Gulf and 215,000 at the Front). Still, sealers took an annual average of 285,000 seals from 1961 to 1965.[21]

By now, an increasing number of individuals, Canadian government scientists, and groups like the Canadian Audubon Society were beginning to worry openly that the harp seals would not survive without stricter controls on the hunt. Before long, as chapter 20 will document, animal rights activists were joining forces with environmentalists to call for an end to this 200-year-old commercial hunt.

20

The Brutes! Killing Markets with Activism

Activists in the 1960s made little headway convincing Newfoundlanders to end the annual hunt for harp seals. Although the Canadian federal government did begin to regulate harvesting and impose quotas, activists rejected this response as inadequate. The anti-sealing campaign began to gain ground in the 1970s as environmental groups like Greenpeace joined forces with animal rights groups like the International Fund for Animal Welfare. The activists were idealistic, imaginative, and daring. TV crews captured them confronting hunters with spray paint in a bid to save a few seal pups. Movie stars joined them to cuddle whitecoats in front of millions of viewers.

As years passed, the activists became better organized and began to raise funds for million-dollar budgets. They filled conventions with snow-white balloons and handed out T-shirts with pictures of teary whitecoats to schoolchildren. Petitions with thousands of pages of signatures calling for the hunt to end were delivered to politicians. Over time, the activists also began to focus less on influencing Canadians and more on disrupting the import of Canadian products. This strategy would eventually produce a dramatic victory when activists succeeded in recasting the seal hunt as an immoral slaughter in the minds of European consumers.

Confronting the Seal Hunt in the 1960s

Before 1950, few outside the sealer communities knew anything about the Canadian seal hunt. This began to change by the late 1950s and early 1960s with the first ethical rumblings among animal welfare activists over the "inhumane" methods of the sealers.

In 1955, Harry Lillie, who'd served as a medical officer on a sealing ship six years before, filmed the Newfoundland hunt and sent copies to humane societies; he then published *The Path through Penguin City* on the brutality of the seal hunt. Many found his firsthand accounts disturbing. Seal pups, he wrote, "were generally killed quickly by a blow on the head, but occasionally I saw men in a hurry just daze them with a kick and cut the little bodies out of the pelts while they lay on their backs still crying."[1]

In 1960, the Newfoundland author Harold Horwood, writing in *Canadian Audubon* magazine, called for a ban on hunting adults, on panning, and on sealing in May, as well as a limit on the number of vessels and size of catches for both Canada and Norway.[2] Then, in 1964, came a documentary by Artek Films with footage of a landsman skinning a seal alive. This was the first anti-sealing film to reach a more global audience, playing on Télévision de Radio-Canada in Quebec and later on German television. After seeing the film, Montreal journalist Peter Lust wrote the article "Murder Island," which was published in over 300 newspapers worldwide.[3] Later, the landsman in the Artek Films documentary made a sworn statement saying the filmmaker had paid him to skin the seal alive—and that this had happened *before* the official opening of the sealing season. In a letter at the beginning of Lust's 1967 book *The Last Seal*, the filmmaker denied these allegations and blamed a camera crew from the Canadian Broadcasting Corporation for staging the skinning of a seal *after* the end of the season.[4] The controversy over the authenticity of the Artek Films documentary didn't really matter in the end, though: the film was the beginning of a global awakening to the "horror" of the seal hunt.

Canada's Ministry of Fisheries received thousands of letters from people outraged by what they'd seen; the federal government responded in October 1964 with new sealing regulations. It required all sealers to obtain a license and restricted the seal-hunting season in the Gulf and at the Front to seven weeks. It set a quota of 50,000 seal pups for the Gulf of St. Lawrence, which effectively ended Norwegian sealing there. It made skinning a seal alive illegal and imposed new standards to ensure that clubs were big enough to kill seals quickly. The government also banned overnight panning (except during storms) and the killing of adult seals in breeding patches.

The effect of these restrictions was not dramatic: the average annual harvest of northwestern Atlantic harp seals fell only slightly from 285,000 in 1961–65 to 280,000 in 1966–70. The introduction of quotas, however, shifted the focus of hunters from seals for leather to seals for furs. The value of Canadian seal leather exports was about equal to seal furs in the 1950s. With improving techniques to keep the fur of the pelts firmly attached to the skin, however, seal fur prices began to rise in the early 1960s. Faced with a quota, whenever possible, sealers hunted seals whose fur was in excellent condition—that is, whitecoats, not ragged jackets (molting seals, 2–5 weeks old).[5]

Many in the growing movement for the humane treatment of animals saw the Canadian government's response to protests as utterly inadequate. Some, most notably Brian Davies of the New Brunswick Society for the Prevention of Cruelty to Animals (SPCA), began to call for an end to the seal hunt. Davies first went to the Gulf hunt in 1965. In 1968, he led 18 observers to the Gulf hunt, including a photographer and a reporter from London's *Daily Mirror*. The following year, he brought a reporter and photographer from *Paris Match*. By the late 1960s, these efforts had stirred up intense critical publicity in Great Britain and France. This publicity in turn disrupted markets, and prices for seal pelts tumbled from 1965 to 1968.[6]

The Canadian government took further measures to make the seal hunt more humane. For the 1967 hunt, it imposed stricter specifications for clubs and banned the gaff (which some sealers would use to hook into live seals). Inspectors were also granted the power to revoke the license of a sealer guilty of inhumane hunting techniques. These new measures upset many sealers. Newfoundland sealers were particularly angry about the ban on the gaff, which many of them saw as vital for their safety—both to hook onto the ice should they fall into the water and to haul men out of the water.[7]

Measures were also taken to ease the commercial pressures on the sealing population. This was difficult for Canada to achieve alone because the Front, where most of the Norwegians hunted, was in international waters. Some efforts at cooperative management did occur—in 1961, for example, Canada and Norway agreed to end their hunts on the 5th of May—but the continuing decline in the number of seals showed them to be inadequate. The international community became especially alarmed by a 1964 survey that found sealers took 85 percent of the pup population at the Front and 53 percent in the Gulf.

In 1966, Italy became the last member state to sign the Harp Seal and Hooded Seal Protocol of the International Commission for the Northwest Atlantic Fisheries (ICNAF), for the first time establishing international regulation of the seal hunt at the Front. The first measures taken were for the 1968 hunt, whose opening date at the Front was changed from 12 March to 22 March to lessen the pressure on the whitecoats. The closing date for the 1968 hunt was also changed, from 30 April to 25 April, to reduce the number of breeder seals caught. These measures did little, however, to halt the decline in the population of northwestern Atlantic harp seals, which, by 1970, was down to between 1.6 and 1.8 million—a drop of well over a million from 1950 (3.3 million). Although far lower than during the heyday of the seal hunt in the mid-1800s, the 1970 figure was somewhat higher than the 1960 one (perhaps reflecting the difficulty of accurately determining herd size).

Confronting the Seal Hunt in the 1970s

At the urging of the Canadian government, the International Commission lowered the quotas for harp seals at the Front three times in the 1970s: to 245,000 (100,000 each for Canadian and Norwegian vessels and 45,000 for landsmen) in 1971; to 150,000 (60,000 each for Canadian and Norwegian vessels and 30,000 for landsmen) in 1972; and, finally, to 127,000 (52,333 for Canadian vessels, 44,667 for Norwegian vessels and 30,000 for landsmen) in 1976. The Canadian government moved to protect the Gulf herd as well, banning the whitecoat hunt there in 1970 and the use of vessels over 65 feet in 1972: decisions that, in effect, switched the large-vessel hunt entirely to the Front, while reserving the Gulf for landsmen hunters.

These measures did little to appease the campaign against the seal hunt, which had gained momentum in 1969 with Brian Davies's founding of the International Fund for Animal Welfare (IFAW), whose mission was to end the commercial seal hunt in Atlantic Canada. Opposition to the hunt was growing within the United States as well. The Marine Mammal Protection Act of 1972 forbade the import of products from nursing marine mammals less than eight years of age, effectively banning the import of whitecoats. Although this had little effect on the Atlantic seal hunt (by 1972, U.S. demand for Canadian seals was negligible), it was another sign of the growing backlash to the hunt.

Yet another sign was the increasing ability of protestors to raise funds. The IFAW was raising over a half million Canadian dollars a year by 1973, the year it hired New York advertising firm McCann-Erickson (whose clients at the time included Coca-Cola) to run its "Stop the Seal Hunt" campaign. The firm's commission of C$100,000 turned out to be a sound investment: the following year, contributions to IFAW exceeded C$800,000.[8]

Cuddly baby whitecoats became ubiquitous: on posters, in pamphlets, as stuffed animals. The accompanying language of "baby seals" butchered in "nurseries" played on parental instincts to protect human babies. The deaths of these weeping beauties tore at the conscience of consumers; many were horrified by graphic footage of pups seemingly skinned alive.[9] Sealers and the governments of both Canada and Newfoundland struggled to counter these images with more businesslike language: "harvest," "rational management," and "traditional product." But the activists were clearly winning the war of words to describe and interpret the "facts" about the hunt. For increasing numbers of consumers, the once brave hunt by selfless providers was now a senseless slaughter of endangered innocents for the whims of the luxury fur market. For millions around the world, swilers were now "killers" and seals now "victims."[10]

For 1976, IFAW contributions climbed to over C$1 million. Brian Davies flew a group of American airline stewardesses by helicopter to that year's hunt—in part to try to debunk the "myth" that the modern-day hunt was dangerous. Greenpeace also traveled to Newfoundland in a storm of publicity after announcing a plan to spray whitecoats with a nontoxic but indelible green dye—and thus destroy much of their commercial value.[11]

Greenpeace Joins the Campaign

The electronic and print media from around the world—NBC News, *Der Stern*, the *Washington Post*, and the Canadian Broadcasting Corporation—pounced on the spectacle of a group of young idealists challenging the hardened folk of Newfoundland. Greenpeace focused more than the IFAW on the environmental threats of the hunt. The ever downward trend was irrefutable in Greenpeace's view: a seal population falling from 20–30 million in the 1700s to 10 million by 1900, 3 million by 1950, 1.5 million by the early 1970s, and 1 million by the mid-1970s.[12]

Although an angry mob of Newfoundlanders confronted the first wave of Greenpeace activists in the spring of 1976, a compromise was soon reached. Greenpeace agreed not to spray seals with green dye, to leave landsmen alone, and to focus the protest on Norwegian "factory ships." In exchange, the locals agreed to allow (or at least not to block) Greenpeace helicopters from reaching the ice floes. Both sides agreed as well to pressure the Canadian government to declare a 200-mile fishing management zone.

When headlines like "Greenpeacers 'Converted' to Sealers' Side" appeared the next day, many Greenpeace supporters became angry. A Toronto Greenpeace group even mailed a bag of crushed Greenpeace buttons to Greenpeace President Robert Hunter in St. Anthony's, New-foundland. But the compromise allowed activists to reach the whelping ice, film the hunt, and later use the footage to stir up a global furor of negative publicity.[13]

The following year, Canada adopted a 200-mile fishing zone, thus becoming solely responsible for management of the Front. The federal government set a quota of 160,000 harp seals in the Gulf and at the Front and 10,000 for native hunters in the Canadian Arctic. It raised the landsmen's quota to 63,000, setting the Canadian fleet's quota at 62,000 and the Norwegian fleet's at 35,000 (11 vessels went to the 1977 hunt: 6 from Canada and 5 from Norway). To focus the hunt on whitecoats and immatures, it limited the large-vessel catch of adult seals to no more than 5 percent of the total catch.[14]

Greenpeace and the IFAW returned to the ice floes in March of 1977. The appearance of French actress Brigitte Bardot—and the photo on the cover of *Paris Match* of her cuddling a whitecoat—sparked worldwide publicity. Recognizing the growing power of this campaign, the Canadian government moved against the activists. It prosecuted Brian Davies for vio-lating sealing regulations by bringing his helicopter too close to the seal hunt; his conviction prevented him from returning to the ice floes in 1978.

To block other protestors from returning to the ice in the spring of 1978, the Canadian government issued an order in council on 26 February 1978, making it illegal for non-sealers to go on the whelping ice without a permit. In a barefaced move, the initial order set the appli-cation deadline for the permits as one week *before* the order was issued. Greenpeace lawyers managed to force the government to waive this impossible deadline. The fight for a handful of permits was nonetheless bitter. Greenpeace members Rex Weyler and Patrick Moore were arrested and charged with "loitering" in a Department of Fisheries office for trying to obtain a permit. In the end, Moore and Weyler did make it to the ice, although Moore was again arrested and charged with violating the Seal Protection Act for interfering with the hunt. His crime: holding a startled whitecoat.[15]

The actions of the Canadian government, far from stemming the rising tide of global protest, served only to arouse media interest and to provoke consumer anger outside of Canada. The IFAW continued to lobby politi-cians and worked to gain media coverage in Europe—especially in Great

Britain—to influence consumers there. The price of a seal pelt, which had gradually risen to C$20–$30 by the mid-1970s, declined sharply as the protest publicity—and shifting consumer demand, which now favored longer-haired furs—made markets skittish. Meanwhile, the IFAW was steadily growing stronger, raising close to C$1.3 million in 1977.[16] Greenpeace also benefited from the campaign to end sealing; some speculated that it was "subsidizing" its antiwhaling campaign with funds it was raising for its anti-sealing campaign.[17]

Another group—the Fund for Animals based in New York—joined the protest in late 1978 by announcing a campaign to organize a tourist boycott of Canada until it banned the seal hunt. By the beginning of 1979, the group had mailed millions of letters throughout the United States. The Fund for Animals tried to disrupt the 1979 hunt in the Gulf (which the Canadian government reopened to one large vessel in 1978 and two large vessels in 1979) by sailing the *Sea Shepherd* to the whelping ice and spraying hundreds of whitecoats with red dye. (The crew included Paul Watson, a founding member of Greenpeace and one of the protestors at the 1976 hunt.) Greenpeace went back to Newfoundland in 1979, too, announcing it would now oppose the landsmen hunt. It managed to spray some green dye on a few whitecoats during one brief trip to the ice on 14 March 1979, but could do little more after the Canadian government revoked its permit.

The Atlantic seal hunt in the 1970s took far fewer harp seals than in previous decades, partly because of the protests and resulting market instabilities, partly because of government quotas, and partly because of the comparatively small numbers of whitecoats. The harvest fell about 50,000 from the 1966–70 average to about 230,000 in 1971, then dropped below 170,000 for every year of the 1970s (with lows in 1972 and 1973 of between 125,000 and 130,000) except for 1975, when sealers took about 175,000.[18] Environmental protestors characterized the lower harvests as a sign, not of better management, but of a species on the brink of extinction. Animal rights groups were upset that whitecoats still constituted nearly 80 percent of the catch. By the beginning of the 1980s, the backlash to the seal hunt was growing even stronger, especially among European consumers.

The Consumer Campaign

Greenpeace continued its campaign of direct action against the seal hunt into the early 1980s. It sailed the *Rainbow Warrior* to the Front in 1981 and to the Gulf in 1982, where officials arrested members for again spraying green dye on seals. The IFAW, meanwhile, was focusing on mail-in and petition campaigns in Europe, North America, and Australia. It bought full-page ads in European newspapers asking readers to write to members of the European Parliament to encourage an import ban on whitecoats and bluebacks (nursing hooded seals). At least 3 million letters and postcards—and perhaps as many as 5 million—descended on the European Parliament.[19]

In March 1982, as members of the European Parliament debated a proposal to ban whitecoat and blueback seal imports, the Canadian delegation scrambled to convince parliamentarians of the economic value, sustainability, and humanity of the seal hunt. It did so amid a throng of protestors waving petitions from school children, handing out baby-white balloons with teary eyes, and showing graphic pictures of sealers bashing pups.

The Canadian countereffort—an approach that included handing out buttons saying, "Save Our Cod, Eat a Seal"—was a complete failure. The vote in the European Parliament was decisive: 160 to 10 to ban whitecoat and blueback seal imports into countries of the European Union. Though nonbinding, it boded ill for the Canadian government's countereffort.

In October 1982, the European Commission agreed with the European Parliament and recommended a temporary ban on the commercial import of harp whitecoats and hooded bluebacks. In something of a stretch, the commission justified its decision by citing the antipornography clause of the GATT, which allows member states to protect public morals. The European Parliament then passed a temporary import ban in November 1982, effective 1 March 1983.[20]

Meanwhile, the Atlantic seal industry seemed to have recovered slightly in the few years before the European import ban. Harp seal harvests exceeded 165,000 in 1980 and 195,000 in 1981. Because, however, Europe was importing almost three-quarters of all Canadian seal pelts at this time, the ban was devastating. The average price of a seal pelt in 1983 fell by half from the previous year—to just C$13 in 1983. Only three vessels—one from Nova Scotia, one from Quebec, and one from Newfoundland—had even bothered to go to the ice in the spring of 1983, and the total harvest was just 30,000 pelts.

The activists kept up the pressure even on this comparatively small hunt. Paul Watson, now belonging to the newly formed Sea Shepherd Conservation Society, sailed the *Sea Shepherd II* to just outside St. John's harbor to block sealers. When, after two weeks, no vessel even attempted to leave the harbor, Watson sailed on to the Gulf, where the Canadian police boarded the *Sea Shepherd II* and arrested him.

The IFAW, now sensing a full victory, pressed on with the campaign on other fronts. In late 1983, it began to lobby consumers and supermarkets in the United Kingdom to boycott Canadian fish products until Canada banned the seal hunt. It distributed over 4 million preprinted postcards urging supermarkets to remove Canadian fish products from their shelves and freezer bins. This yielded quick results. In early 1984, Britain's biggest supermarket chain, Tesco, announced it would not buy any more Canadian fish products until the hunt ended; the Safeway supermarket chain quickly followed suit. The IFAW then extended its boycott campaign to the United States—which, at this point, accounted for 80 percent of Canada's fish exports—printing 5 million boycott postcards for American consumers, with targets like McDonald's and Burger King.

This threat to Canada's fishing industry ignited calls within Canada to concede defeat and end the seal hunt. Sealers—from Canada's poor eastern provinces—wielded little influence on federal politics. Moreover, the sealing industry was only a small part of the economy, even for Newfoundland, and, although many Newfoundlanders were angry with the activists, few wanted to risk a boycott of Canadian fish products. In 1982, the landsmen broke ranks with the sealing fleets and formed the Canadian Sealers' Association, which called for a moratorium on the hunting of whitecoats. This was a safe way to diffuse the global protest without significantly hurting the income of landsmen, most of whom hunted older seals—beaters, bedlamers (immatures from 1 to 4 years of age), and adults.

The Canadian government stood firm, however. As part of its ongoing effort to counter the image of a slaughter of "babies," the Fisheries Department explained in the *Toronto Star* of 15 February 1984: "A ragged jacket is like a 20-year old leaving home, and by the time it reaches a year old, it's been through a couple of jobs and a divorce."[21] Going on the offensive in an article in the *New York Times* on 10 March, Fisheries Minister Pierre De Bané called the organizers of the boycott of Canadian fish products "blackmailers," "liars," and "fascists."[22]

Although the large-vessel seal hunt was still technically legal in 1984, no vessels went to the ice. For the first time in centuries, the hunt was left entirely to landsmen. This was the case in 1985 and 1986 as well, after Europe renewed the import ban in 1985. When, in March 1987, two vessels ventured out to hunt beaters and adult seals with rifles, because of heavy ice, they landed fewer than 3,100 pelts.[23] This move, however, sparked an angry reaction from activists. The three core groups—Greenpeace, IFAW, and the Sea Shepherd Society—didn't have time to organize a protest of the 1987 hunt. But it was clear that, if the hunt were to continue, the protestors would be back in force the following spring.

Banning the Hunt for Whitecoats

The population of northwest Atlantic harp seals was now rebounding. By the mid-1980s, there were about 2 million harp seals, with mothers giving birth to about 500,000 whitecoats, and the population was increasing with each season.[24] Clearly, harp seals were no longer endangered, much less under threat of extinction.

Nevertheless, at the end of 1987, the Canadian government decided to ban the hunting of whitecoats and bluebacks and the hunting of other seals from large offshore vessels (over 65 feet or 19.8 meters in length). It made these decisions, Newfoundland Member of Parliament John Crosbie would later admit, in large part because of the threat of an IFAW-organized boycott of Canadian fish products, which potentially had a far greater economic impact than the loss of export revenues from a few offshore vessels.[25] The government still allowed landsmen to shoot seals from small boats. The 1983 annual quota of 186,000 remained in place, but it was far beyond the landsmen's capacity to reach and the Canadian harvest of harp seals averaged only 51,000 a year from 1983 to 1995.

For years, activists celebrated the effective end of Canada's commercial seal industry as a triumph of decency and ecology over needless luxury.[26] In a 1989 autobiographical reflection, Brian Davies called it "a victory for Canada"[27]—one applauded by the Sea Shepherd Conservation Society, the International Fund for Animal Welfare, and Greenpeace, who now shifted their resources to other causes. But, as chapter 21 will show, th' battle wasn't over yet.

21

Hunting Beaters for Globalizing Markets

In the mid-1990s, the Canadian government—with the population of harp seals now over 5 million—took steps to expand the fleet of small vessels hunting "mature" seals. It kept the ban on hunting 6- to 12-day-old whitecoats in place, but quietly raised harvesting quotas for beater seals (2–12 weeks old), providing direct and indirect subsidies to small-vessel fishermen and seal-processing facilities to revive commercial interest. Before long, prices, markets, and profits for seal pelts were rising. Today, this small-vessel hunt is the biggest in over a half century, turning what activists once saw as a lesson in how to use the media to alter consumer consciousness into a lesson on the power of governments and entrepreneurs to keep consumption rising in an era of globalizing markets.

The resurgence of the seal hunt came as a slow surprise to many anti-sealing activists. Many are now back campaigning, furious at what they see as a deceptive reopening of the world's greatest annual slaughter of marine mammals. Once again, Greenpeace and the International Fund for Animal Welfare are cooperating in an anti-sealing coalition, along with more than 50 other groups. The celebrities are back as well, from actor Martin Sheen to musician Paul McCartney. The campaign is not much different from the one in the 1970s and 1980s—with media stunts at the spring hunt, footage of sealing, petitions for politicians, efforts to disrupt markets, and boycotts of Canadian seafood.

But, this time, activists have made little headway influencing governments or consumers. The total harp seal catch remains high—with the decrease in the 2007 quota (by 55,000 seals) a response to poor ice conditions, not pressure from environmentalists or animal rights activists. Why is today's campaign less successful? A variety of factors seem to be at play. Some consumers—even in Europe—are now wearing furs to show their antagonism to environmental preaching. More consumers seem to see today's seal hunt as sustainable and humane. The Canadian government is also far more adept at handling the mainstream media: public relations teams have countered the activists with images of salt-of-the-earth fishermen, statistics showing an abundance of seals and a shortage of cod, and the language of sustainable and humane management.

In addition, unlike the activist coalition of the 1970s and 1980s, today's coalition is diffuse and spread thin: many groups are tiny outfits; the major groups of the past—in particular, Greenpeace and the International Fund for Animal Welfare—are now complex organizations pursuing multiple interests. Coordinating a campaign is more difficult, and for some campaigns, such as the boycott of Canadian seafood, sharp differences can emerge over appropriate strategies. The globalization of activism and governments' greater skill at turning idealism into cynicism are also making it much harder for activists to use the media as a free pathway to change the "consciousness" of consumers.

The most significant factor countering the influence of today's anti-sealing activists, however, is the globalization of markets for Canadian seal furs. Activists continue to make gains in Europe—for example, in 2007, Belgium, Italy, and Luxembourg banned the import of *all* seal products, not just whitecoats. But these markets are tiny. On the other hand, in Russia and China, which now import 90 percent of processed harp seal furs (treated and dyed), activists have made little progress influencing either consumers or governments. To take a closer look at this activist campaign, we need to step back to the mid-1990s and review how the Canadian government managed to revive the commercial hunt for harp seals.

Awakening the Industry

The hunt for harp seals since 1996 is not, the Canadian government stresses, like the commercial hunt of the past. Under the Marine Mammal Regulations, it remains illegal to harvest, trade, or sell whitecoats, ensuring a more "humane" hunt. Quotas on the total allowable harvest are carefully set to ensure that the hunt is "sustainable." Moreover, sealers in vessels over 65 feet (19.8 meters) are permitted to participate only as "collectors" so that out-of-work fishermen with small vessels may earn income from the hunt.

Back in 1995, the government justified raising the quota on the total allowable catch to protect the collapsing cod stocks on the Grand Banks off Newfoundland from the "exploding" population of harp seals. "There is only one major player fishing that stock," Canadian Fisheries Minister Brian Tobin explained. "And his first name is harp, and his second name is seal."[1] Indeed, Tobin estimated that harp seals were consuming a billion Atlantic cod every year. Thus, given the abundance of seals and the poverty of the Atlantic Canadian fishing communities, expanding the commercial harvest was the only rational and responsible course to take.[2]

Following this logic, the Canadian government raised the quota on harp seals for the 1996 hunting season to 250,000. It also began to provide millions of dollars in direct and indirect subsidies to revive the industry (aided by rising pelt prices). The turnaround was immediate. The harp seal catch for 1996 jumped to over 240,000, rising again in 1998 to over 280,000, before falling off to just under 245,000 in 1999. Although low pelt prices, high operating costs, and poor ice conditions combined to reduce the 2000 harvest by more than half—to only 91,600—the catch rebounded to over 226,000 in 2001.[3]

Expanding Markets

Over this time, the Canadian sealing industry was working hard to expand markets for harp seal pelts. The Canadian Seal Industry Development Council, for example, went on tour in the late 1990s to convince the fur industry to reintroduce seal pelts. This was—and remains—a formidable challenge: 80 percent of Canadian furs are sold in the United States, where the 1972 Marine Mammal Protection Act prohibits the sale of beater pelts.

Seal furs were slowly coming back into fashion internationally, however, in part because demand for real fur coats began to rise in the late 1990s. Supermodels, who once declared they'd "rather go naked than wear furs," began to arrive at fashion shows draped in real furs. Retail sales of furs in 2002, for example, rose over 10 percent globally, making it the "best-performing luxury item" of the year. Furs were becoming trendy among younger consumers: the average age of customers in 2002 fell from 49 to 35. Even in the United Kingdom, once one of the world's most hostile markets to furs, retail sales of fur and fur accessories rose by 35 percent in 2002. Some young customers—with enough irony to make even a hardened swiler smile—were telling marketers they were "tired of being preached at" and felt a sense of rebellion with the purchase.[4]

With the market for harp seals surging, the Canadian government ended direct subsidies to sealing in 2001.[5] The total harp seal catch in 2002—312,367—broke the 300,000 mark for the first time since the 1960s. (The Canadian government allowed sealers to exceed that year's quota because of the exceptionally low harvest in 2000.) With international prices for harp pelts holding strong, and with seal meat and other seal products adding another million dollars, the total landed value of harp seals for 2002 was around C$21 million—nearly four times higher than the previous year (C$5.5 million).[6]

To further expand the now-flourishing sealing industry, the Canadian government raised the triennial quota for harp seals to 975,000 over the next three years (2003 through 2005). Canadian sealers took 283,500 harp seals in 2003, although, with a dip in pelt prices, the landed value fell to just C$13 million. In 2004, the Canadian government issued over 15,000 seal licenses (including professional, assistant, and personal use) and, hunting under favorable weather conditions, sealers took 366,000 harp seals—the most in 50 years or so.[7] The landed value of seals in 2004 was almost C$16.5 million: higher than 2003, though lower than 2002.[8]

The pelts of seals under the age of 3 months accounted for most of this harvest, in part reflecting market demand. The average price of harp seal pelts varies widely depending on the age of the seal, and harp seals under a year old are the most valuable on the world market. Thus, in 2004, the average landed value of a ragged jacket pelt (from a molting seal, 2–5 weeks old) was C$16, and that of a beater pelt (from a fully molted seal, less than 3 months old) was C$48, whereas the average landed value of an adult pelt was only C$7.[9] In 2004, the market for seal meat remained small, as it did for seal oil, although the growing demand for oils rich in Omega 3 has the potential to change that for the better.[10]

As in the 1980s, the main consumer markets for harp seal pelts are overseas. About two-thirds are exported raw (untreated). Norway was the primary buyer of raw pelts in 2005, importing nearly 130,000; Greenland was second, at 46,600, followed by Finland, at just over 19,000. The rest of the pelts are "dressed" (treated and dyed) in processing plants in Canada. About 90 percent of these in 2005 were sold to brokers in Russia and China. The demand here is rising quickly. Dion Dakins of the Barry Group of Companies, which runs two fur-dressing plants in Newfoundland, explains the trend: "Russia is the No. 1 customer, with China coming on with insatiable demand."[11]

A Sustainable and Humane Harvest?

According to Fisheries and Oceans Canada, the population of Northwest Atlantic harp seals is stable and growing despite the now-high catches of beater seals: the total seal population was estimated at 5.2 million in 1999, 5.5 million in 2000, and 5.9 million in 2004. The total allowable catch for 2005 was set at 319,500. Even though the 2005 harvest was lower than 2004, the total landed value was about the same (C$16.5 million) because of higher pelt prices. The average price of pelts in 2005 was about C$52—up 18 percent from the 2004 average price.[12]

The Canadian government raised the allowable catch of harp seals slightly for 2006, to 325,000—with another 10,000 set aside for First Nations communities. This was in place for 2007 as well, but after so many pups drowned in the broken ice of the Gulf, the government decided to lower it to 270,000 just as the spring hunt was starting.[13] The northwest Atlantic herd is now three times larger than it was in the 1970s, and, although higher quotas in recent years will likely reduce its numbers slightly once the remaining beaters mature into breeder seals, it is at no risk of becoming endangered.

Some government officials go even further, reversing the arguments of many environmental activists, and characterizing the hunt as necessary for a sustainable ecosystem. "It's not just about revenue," explains Fisheries Minister Loyola Hearn in 2006. "It's about sustainability of a herd. What happens if we don't ensure there's a proper balance in nature? What happens if the herd continues to grow to the point where . . . they have self-destructed?"[14]

New regulations strive to ensure that sealers obtain the full commercial benefits of a seal carcass (and thus help avoid waste). As of 2003, sealers must land the entire carcass or pelt—to prevent them from harvesting a seal only for a portion (such as an organ). Under a Canadian federal apprenticeship, they must also learn how to kill swiftly and efficiently. When using a club or *hakapik*, for example, they must "strike the seal on the forehead until its skull has been crushed."[15] (More sealers, however, now shoot seals from a boat since beaters are more agile and wary than whitecoats.) A 2003 amendment to the Marine Mammal Regulations requires hunters to test a seal's eyes for a blinking reflex by touching them to ensure brain death before skinning and bleeding begins. (Previously, hunters were allowed to check the skull to confirm death.) The federal government claims these regulations ensure more than ever before a "quick and humane dispatch of animals."[16] The Canadian

Veterinary Medical Association concurs, concluding in a recent study: "currently, the large majority of seals taken during this hunt are killed in an acceptably humane manner."[17]

Such regulations, like others of the last four decades, have done little to appease environmental and animal rights activists. Many are once again outraged, and a global campaign to end the commercial seal hunt is again swirling off the Newfoundland coast.

The Anti-Sealing Coalition

Over 50 activist groups now oppose Canada's seal hunt. Once again, these activists are trying to frame the debate in language that challenges consumers to consider the moral and environmental consequences of the hunt. Groups like the Humane Society of the United States call it "the largest commercial slaughter of marine mammals on the planet." Groups like the International Fund for Animal Welfare charge that the "annual baby seal hunt" is still "unacceptably cruel."[18] The Green Party of Canada agrees, opposing the hunt because of its "extreme brutality." Actor Martin Sheen, speaking in his *West Wing* presidential voice on behalf of the Sea Shepherd Conservation Society, calls for an end to the "annual ritual of blood and slaughter of the innocents." Former Beatle Paul McCartney, speaking to the press during a protest in the Gulf of St. Lawrence in March 2006, called the hunt "heartbreaking," and "a stain on the character of the Canadian people." The Sea Shepherd Conservation Society talks about the Canadian government's "kill quota" and its "death sentence" for millions of baby harp seals.[19]

The IFAW estimates that hunters killed almost one-third of the harp seals born in 2005 and that, even though these seals were not whitecoats, they were still young and defenseless pups. According to the Humane Society of the United States, many of the seals were as young as 12 days old, while nearly all—about 95 percent from 2000 to 2005—were under 12 weeks old.[20]

All of these groups see the clubbing of beater seals as inhumane. Nor, they are quick to point out, is shooting them a humane alternative: many of these seals suffer before sealers arrive from the boat or when they slip away to die. Groups like the IFAW and Sea Shepherd Society further claim that hunters are still skinning seals alive. Indeed, the IFAW charges that, despite the new regulations, few sealers actually check for a blinking reflex before they begin skinning; the Sea Shepherd Conservation Society puts the proportion of seals skinned alive at 42 percent.[21]

For some activists, the current hunt is not just barbaric, but unsustainable, too. The IFAW argues that the resurgence of sealing since the mid-1990s is a result of short-term political decisions rather than long-term conservation, that the number of seals "culled" is neither "biologically sustainable" nor "scientifically justifiable." Moreover, the estimates of ecological impacts based on "quotas" and "landed catch" are misleading; they do not adequately account for seals shot and lost or discarded because of damage to the pelt.[22]

Greenpeace also opposes the hunt on environmental grounds. It calls the 2003–2005 quotas "irresponsible" and based on science that is "inaccurate, incomplete and out-of-date." The quotas, it argues, do not sufficiently account for many complex factors that threaten the vitality of the Northwest Atlantic seal herds: "struck and lost rates," illegal hunting, seals killed as bycatch in fishing nets, the hunt of the same seal herds in Greenland, and uncertainty in predicting seal birthrates 10–15 years into the future, to name just a few. Nor do they account for the potential of climate change to disrupt whelping ice conditions (as happened in 2007) and thus threaten the survival of the entire seal population.[23]

In January 2005, a coalition of activist groups against the seal hunt also began a Canadian seafood boycott campaign, targeting principally U.S. and Canadian retail stores and restaurants, in particular, North America's largest seafood chain, Red Lobster, with over 600 restaurants. Included in the coalition are the Animal Alliance of Canada, the Humane Society of the United States, the World Society for the Protection of Animals, Environment Voters, the Sea Shepherd Conservation Society, and People for the Ethical Treatment of Animals (but not Greenpeace or the International Fund for Animal Welfare, even though both are members of the Unified Opposition against the hunt). The goal of the boycott is "to take all the profits out of sealing, and for every year the seal hunt continues to directly cost those who kill the seals 50 to 100 times more than what they earn from the commercial seal hunt."[24]

The coalition hopes to accomplish its goal by hurting the Canadian seafood industry, which grosses some C$3 billion a year, almost three-quarters from exports to the United States. Thus far, however, only a handful of buyers accounting for a tiny portion of Canadian seafood—ones like Wild Oats Markets, Kimpton Hotels and Restaurants, and Original Fish Legal Sea Foods—have joined the boycott.

Missing the Consumers

Many factors explain why today's larger anti-sealing movement is not managing to influence consumers as much as the smaller movement did in the 1970s and 1980s. Because their groups are diffuse, today's activists scatter their energies, creating more potential for contradictory messages, and increasing the chances for mistakes that discredit their campaigns. Their tactics are no longer novel news items; far more individuals and groups are competing for that minute or so of daily protest fame. Moreover, the sealers and Canadian government are far more adept at constructing counterimages and counter-language in the media. And, perhaps immune to media horrors, perhaps rebelling against the moralistic tone of some campaigners, many consumers are simply harder to persuade.

Greenpeace and the International Fund for Animal Welfare, the main activist groups of the past, are now larger and more complex organizations, which, at first glance, might seem to suggest a greater ability to challenge the hunt. But these groups are now less confrontational than they were in the 1970s and 1980s, and opposing the seal hunt is no longer so central to their mandates.

A glance at the history of IFAW over the last three decades reveals a significant shift in its organizational structure, focus, and tactics. In 1979, the year Brian Davies moved IFAW headquarters from Fredericton, New Brunswick, to Cape Cod, Massachusetts, it had a staff of just 7; by 1997, when Frederick O'Regan took over from Davies as president, the number of employees in the Cape Cod office had risen to 70. Today its headquarters team includes more than 100 scientists, administrators, and specialists, with an annual payroll of over $3 million.

Worldwide, the IFAW now has over 200 employees and 13 offices. Its mandate both to end cruelty toward animals and to protect species has motivated campaigns for whales, elephants, seals, cats, and dogs, as well as a campaign to end the illegal trade in wildlife. In addition, the fund employs a rapid-response team to assist with the cleanup of oil spills and assist animals in distress. Its diverse branches employ diverse tactics, many of them practical, to achieve change. IFAW scientists, for example, are cooperating with Cornell University to develop an acoustic buoy to help detect whales in busy shipping lanes. The IFAW is also paying local lobstermen to replace floating ropes with sinking ropes to avoid snaring passing whales.[25]

Even as it pursues its broader mandate, the IFAW remains a leader in the anti-sealing coalition. This coalition *is* achieving some "victories." It has managed, as noted above, to advance efforts in Europe to ban the import of all seal products, not just whitecoats, with Belgium, Italy, and Luxembourg imposing full bans by early 2007, if only temporarily. And, even though the European Parliament ultimately rejected a call in March 2007 for a full ban for all 27 members of the European Union, the European Commission decided to investigate the matter further.

But these "victories" are not equal to the ones in the 1980s. In China and Russia, which now import 90 percent of Canada's dressed beater pelts, cultural and political factors are severely limiting the ability of activists to influence governments and consumers. "Our markets are in Russia and China," Canadian exporter Dion Dakins explains, "and they couldn't care less about Paul McCartney."[26] His comment nicely captures the toughest challenge for today's anti-sealing activists struggling to alter the consciousness of consumers—the globalization of markets.

22

The Globalization of Slippery Markets

The history of the Atlantic sealing industry from the mid-eighteenth to mid-twentieth centuries is typical of many renewable "resource" histories over this time. Generations of ordinary men went through harrowing hardships on the ice floes to earn a pittance from the sealing shipowners. The ice and sea took the lives of over 1,000 of these swilers—some, like the men of the *Newfoundland* in 1914, dying in a nightmare of agony. Still, every year, thousands of veterans would return and hundreds of boys would go to the ice to become men. Going to the ice was a way of life, the stuff of song and legend for the sealing townsfolk. With spring hunts in the mid-1800s often taking over a half million pups, swilers worried about storms on the horizon, about ice shifting underfoot, even about fate and spirits—but not about sustainable yields. However cruel to men, the ice protected whitecoat pups, keeping enough predators away, so the sealers reckoned, to ensure an abundance of future breeder seals.

But, as with so many other renewable resources—cod and whales, to name just two—advances in technology along with rising consumption put greater and greater demands on the ability of the harp seal population to renew itself, until it simply could not sustain the increasingly efficient hunts. Owners replaced smaller wooden sailing ships with bigger ones, they replaced sail with steam power, and, finally, they replaced wooden steamers with steel icebreakers, on which they then put refrigeration, radar, and search helicopters. Before long, the "hunt" became a "harvest," with captains filling holds from an ever-smaller stock of whitecoats.

Global crises—the Great Depression and the World Wars—eased some of the commercial demands on the seal herds. But the statistical trend toward smaller herds of northeast Atlantic harp seals was unmistakable: their numbers fell from as many as 20–30 million in the 1700s, to perhaps 10 million by 1900, and to about 3 million by 1950. And even though the Canadian and Newfoundland governments imposed more and more regulations, the number of harp seals kept falling—dropping below 2 million by the early 1970s.

We'll never know whether the Canadian and Newfoundland governments could have, by themselves, managed the population of Atlantic harp seals to avoid a collapse. But eastern Canada's history of the "management" of the northern cod does not inspire confidence in their ability to do so. The matter, in any event, was taken out of their hands when, in the early 1960s, a small group of environmental and animal rights activists launched a campaign that culminated in a European import ban in 1983—and the effective end to the hunt for whitecoats. For the next few years, without buyers or decent prices, hardly any commercial sealers even bothered to go to the Gulf of St. Lawrence or the Front, off Labrador and Newfoundland, and when Canada officially closed the commercial hunt for whitecoats in 1987, many saw it as a sensible decision to repair Canada's image and avoid any boycott of the more lucrative Canadian seafood industry.

How did this small group of activists manage to pull off such a change in global consumption patterns? Why did this happen for whitecoats and not, say, for cod, pigs, or cows? The answer reveals some of the complexity of how and why ecological shadows of consumption shift, intensify, and recede. With daring stunts and celebrities in tow, a small network of committed activists drew media crews from around the world to the hunt. Pictures of the clubbing and skinning of teary-eyed, pleading whitecoat pups—and the sounds of their wailing mothers in blood-red nurseries—shocked the conscience of millions of consumers in North America and western Europe. Donations to organizations like Greenpeace and the International Fund for Animal Welfare began to pour in. More money allowed more elaborate campaigns: with helicopters to the ice, free balloons and T-shirts at rallies, and trips to schools and government offices throughout Europe.

Yet the adorability and accessibility of seal pups only partly explains the power of this campaign to alter the shadow effects of European consumption. Calves and baby pigs are cute, too, and there is plenty of slaughterhouse footage around that's just as disturbing as any of the seal hunt. But ethical and emotional appeals seem to only go so far in transforming patterns of consumption. Many who are willing to forgo a onetime purchase of a luxury item like a fur coat are far less willing to give up meals of hamburgers and bacon.

Anti-sealing activists from the 1960s to the 1980s confronted far less powerful economic and political forces than today's animal rights groups do. Sealing was a small part of the Newfoundland economy and a tiny portion of Canada's. Most sealers and government officials saw the activists more as a nuisance than as a threat to the industry. The result was a response that, in the dry words of the 1986 Report of the Royal Commission on Seals and Sealing in Canada, "can be summarized as ineffective," with the government underestimating the activists and failing "to respond with an effective counter-offensive."[1]

A decade later, however, having learned from this defeat, the pro-sealing lobby was ready to launch a more effective countercampaign. In the mid-1990s, the Canadian government began to increase the allow-able catch, provide subsidies, and seek out new markets. This was *not*, officials stressed, a reopening of the large-scale hunt for whitecoats. It was still illegal to kill whitecoats, and only small vessels and local sealers could take mature seals—an essential culling anyway to protect endangered cod from an exploding population of harp seals. Furthermore, new regulations ensured that the hunt would be both humane and sustainable.

As Canada was launching its countercampaign, the anti-sealing activists of old were working on a host of new issues: whales, elephants, deforestation, climate change, to list just a few. The resurgence of the hunt, which in 1996 landed over 240,000 harp seals—75,000 higher than the annual average from 1972 to 1982—took them aback. Regrouping into a broad coalition, the activists argued that the new language of sealing was loaded with deceit. The hunt was not about protecting cod: harp seals ate hardly any commercial cod. It was not about preserving indigenous or local communities, but about the profits from exporting hundreds of thousands of pelts. Nor was it a hunt for mature seals. Beaters were commonly only a few weeks—and sometimes just a few days—older than whitecoats, and were certainly not, in the words of Assistant Deputy Minister David Bevan of Fisheries and Oceans Canada in 2005, "fully mature, independent animals"—harp seals didn't even reach sexual maturity until 5–6 years of age.[2] And, more important, it was *not* sustainable. Quotas only counted landed seals; they ignored high shoot-and-sink rates and took no account of deteriorating birthing conditions as a result of climate change.

Today's campaigns against the seal hunt resemble the past ones—at least on the surface—with activists and celebrities bearing witness at the spring hunts, with footage of sealers bashing pups, with efforts to undercut consumer prices for pelts, and with boycotts of Canadian seafood. But today's campaigners are confronting a far more effective pro-sealing lobby—one whose counterimages, statistics, and language of sustainability are able to keep markets calm. The coalition of anti-sealing activists is also more fragmented and less effective. Past leaders—Greenpeace and the International Fund for Animal Welfare—now manage corporate budgets and balance numerous crosscutting campaigns. More than 50 smaller anti-sealing groups are now competing for media coverage and donations—and all the groups, large and small, are competing for public attention in a world where more activists are campaigning on more issues in more locales.

The current campaign to end the seal hunt is making some headway in Europe, where, for example, Belgium extended the ban on whitecoat imports to all seal products in 2007. But, with the main markets for processed seal pelts now in countries like Russia and China, bans in Europe no longer pose a threat to the sealing industry. For the first time since the 1960s, sealers took 300,000 harp seals in 2002, then almost 370,000 in 2004, the largest hunt of the last half century. The 2007 allowable catch of 270,000 harp seals was lower than for previous years. But activists could not take credit. It simply reflected the ice conditions that spring, a reason for fluctuating seal catches since the beginning of commercial hunts in the eighteenth century.

The International Fund for Animal Welfare continues to label Canada's ban on hunting whitecoats a "fragile victory."[3] Will this moral victory withstand the rationality and efficiency of globalizing markets? Whitecoats, as swilers have known for centuries, are easier to "collect" than beaters or adults, and when not collected have higher natural mortality rates than beaters: gales, thin ice, and shifting floes can plunge tens of thousands into the cold water, where they haven't enough blubber to survive. Also, absent an outright ban, their pelts tend to fetch the highest prices. Will emerging markets for luxury fur in countries like China and Russia overcome the moral blockade on whitecoats? The honest answer is "maybe," as activists, corporations, and governments struggle to frame what is environmentally right and ethically wrong for people to consume.

Conclusion
Transforming Global Consumption

23

The Illusions of Environmentalism

The ecological shadows of consumption are continually shifting—moving from place to place, advancing, and receding. A rising global population and rising rates of personal consumption are causing these shifts, as are the globalizing pressures of corporations, trade, and financing, the values of new generations of consumers, and the consequences of technological and scientific "advances." What are the impacts of these shifting shadows on the global environment? Looking at the consumption of products like automobiles, refrigerators, and beef over several generations, we find that the global environmental impacts of these—and a wide range of other—consumer goods have intensified even as environmentalism has grown stronger, a finding that suggests an urgent need to reform political and personal decision making.

Why is this happening? This chapter argues that environmentalism has failed to slow the ways that producing, using, and replacing consumer goods deflect ecological costs into distant places and future generations. Consumption, interacting with political and economic structures, continues to deflect these costs into ecosystems with less capacity and onto people with less power to adapt to them, and it is doing so at a quickening pace. Moreover, even as the globalization of environmentalism reduces the per unit impacts of consumption and imposes global controls through international agreements, the economic forces of globalization are casting and lengthening ecological shadows. Economic globalization is also diminishing the capacity of activists and states to influence the direction, speed, and intensity of the environmental consequences of consumption—a key finding in the case of the harp seal—which helps explain why so many ecosystems continue to slide into crisis even as so many of us celebrate the progress of environmentalism. The nature of that progress has much to tell us about how environmentalism is altering the global political economy of consumption.

The Progress of Environmentalism

A strengthening of environmentalism over the last half century has done much to prod the global political economy in new directions.[1] Hundreds of international agreements now aim to protect ecosystems from the consequences of rising consumption, with controls on trading in endangered species, dumping hazardous waste, and emitting pollutants. At the same time, governments everywhere are adjusting domestic policies to mitigate the environmental impacts of economic growth. Examples are easy to find. Rules for storing chemicals and disposing of waste are becoming more rigorous. Standards for auto emissions continue to tighten. Targets for energy efficiency are getting higher. Labeling programs to alert consumers of risks are becoming more common. And the rules for conserving parklands are becoming more exacting.

Governmental capacity in the developing world to implement environmental policies is strengthening as well. Donors like Japan are providing bilateral grants and technical support to assist with training staff. Lenders like the World Bank are giving funds and policy advice, and agencies, as in the case of the Global Environment Facility (GEF), are financing the additional costs of meeting global environmental commitments. Since its founding in 1991, and acting through the World Bank, the United Nations Development Programme, and the United Nations Environment Programme, the GEF has supported over 1,950 projects in 160 developing and transitional countries with more than $7 billion in grants and with another $28 billion in cofinancing.

Just about every multinational corporation is also pursuing a sustainability policy, commonly under the mantle of corporate social responsibility, with some now investing considerable sums in environmental research and technologies. Many are now following voluntary codes of conduct, such as the chemical industry's "Responsible Care" code, and have joined voluntary labeling programs to inform consumers of "sustainable" products. A few, such as the Swedish firm Electrolux, are actively working with suppliers, producers, users, and recyclers in developing countries to help them improve their environmental performance. Increasing numbers of corporations are also joining initiatives like the UN's Global Compact on human rights, labor standards, environment, and corruption.

The capacity of environmental activists continues to increase as well.[2] Hundreds of thousands of diverse groups—large and small—have formed networks advocating for change, many repeating messages and images to embed new meanings and emotions into the public psyche, recasting for consumers, for example, the "hunting" or "harvesting" of seals as senseless slaughter. Such changes are not a matter of activists, however, simply "educating" passive consumers. Advertisers, scientists, government officials, and business leaders are all contesting this education, and the crisscrossing of truths, word maps, and stories creates various outcomes. In some cases, a word with a positive ecological meaning (wetland) has replaced one with a negative common meaning (swamp). In other cases, corporate phrases like "unleaded gasoline" have become part of the consumer vocabulary, leaving the impression that firms are doing consumers a favor. In still other cases, environmentalism has done little to influence the language—and thus also the understanding—of average consumers, as in the case of car "accidents."

The interests and objectives of nongovernmental campaigners are far ranging, from villagers organizing to protect a patch of land in Nicaragua to celebrity advocates like Ralph Nader and Al Gore speaking to a national or global audience. Strategies vary considerably as well, from the "culture jamming" of Adbusters, which runs spoof and counter-ads to get people *not* to buy products, to the direct environmental action of the Green Belt Movement in Kenya, which, under the leadership of Wangari Maathai (winner of the 2004 Nobel Peace Prize) has planted millions of trees in Africa. In this throng of environmentalists, however, a few groups stand out as major multinational players. Thus Greenpeace International, with offices in dozens of countries, has millions of supporters worldwide. Friends of the Earth, with about 1.5 million members and supporters, has evolved into the world's largest network of grassroots organizations, with over 70 national groups and over 5,000 local ones. And the WWF (World Wildlife Fund / World Wide Fund for Nature), with close to 5 million regular supporters worldwide, operates in over 100 countries, employing nearly 4,000 people and funding more than 2,000 conservation projects.

Increasingly, organizations like the WWF are partnering with companies as well. A few examples show the diversity of these arrangements. WWF-Sweden is working with the multinational food firm Tetra Pak on responsible wood purchases and climate change policies, and WWF-India with the Austrian crystal firm Swarovski to establish a wetlands visitor center in the Keoladeo National Park in northeastern India. The WWF is also working with the Finnish state forest company Metsähallitus to protect 55,000 hectares (136,000 acres) of old-growth forests in northern Finland. The WWF has Climate Saver deals with the

Danish pharmaceutical firm Novo Nordisk and 10 other firms as well—including multinationals like IBM, Johnson & Johnson, Nike, and Polaroid. Together, these 11 firms hope to cut their carbon dioxide emissions 10 percent by 2010—or some 10 million metric tons per year, roughly equal to taking 2 million cars off the roads.[3]

Branches of many other activist groups—even the once confrontational Greenpeace—are now cooperating more with states and corporations. The labeling programs of nonprofit NGOs like the Forest Stewardship Council (FSC) and the Marine Stewardship Council (MSC) demonstrate some of the ways these activists are working to influence markets and consumption. The Forest Stewardship Council was founded in 1993 by various stakeholders, including NGOs like Greenpeace, the WWF, and Friends of the Earth, retailers like IKEA, as well as various forest firms and indigenous peoples. The council monitors and accredits organizations to certify wood that meets its principles and criteria for forest stewardship. The FSC logo—now the world's most recognized in the field of sustainable forest management—provides a "credible guarantee" to the consumer "that the product comes from a well-managed forest." Over the last decade alone, more than 50 million hectares (124 million acres) in over 60 countries met FSC standards, and businesses such as Home Depot—which now sells more FSC-certified wood than any other retailer in North America—now rely on the Forest Stewardship Council to assist with purchasing.[4]

The Marine Stewardship Council, like the FSC, uses its logo to influence markets, retailers, and consumers. Founded in 1996 by the WWF and the Unilever food conglomerate to promote responsible and sustainable fisheries, the MSC has since expanded its reach and impact, recently certifying, for example, a cod fishery in the North Pacific and mackerel icefish in Australia as "sustainable." The potential of logos like the MSC's to influence consumption patterns is much greater when retailers agree to carry only certified products. The pledge in 2006 by Wal-Mart—the world's biggest retailer—to purchase wild seafood only from fisheries meeting MSC standards could well encourage other major retailers to follow suit (assuming the pledge is honored), with the potential for MSC seafood to gain a significant share of the global market.[5]

Corporate and consumer responses to the strengthening of environmentalism are opening and expanding many other markets as well. Thus organic agriculture has expanded to over 30 million hectares (74 million acres) of farmland, up from just 5 million hectares (12.4 million acres) in 2005. Thus, too, sales of more energy-efficient products have risen steadily over the last decade, a trend that may even accelerate as governments impose stricter environmental regulations, consumers put more value on conservation, and corporations compete for these markets. Eliminating even seemingly small inefficiencies, such as the electricity wasted by electronic appliances in standby mode, can save considerable resources. In the course of a day, for example, a typical microwave oven uses more energy to power its timer in standby mode than to heat or cook food. Indeed, standby mode now accounts for between 5 and 13 percent of energy consumption for an average household in the First World. More energy-efficient designs, some researchers now argue, could cut this waste by about three-quarters without any loss in performance or convenience.[6] Seeing such waste, a few governments, such as California's in 2006, are now beginning to impose mandatory limits on the amount of energy electronic goods can use during standby mode.

As part of this strengthening of environmentalism, consumers are shifting their choices and practices. Young travelers, for example, now set off to explore the world as ecotourists; some are even "WWOOFing," volunteering to work abroad with World Wide Opportunities on Organic Farms. Others are planting urban gardens on rooftops, composting organic waste, and searching supermarkets for food grown locally. Still others are designing buildings to use more solar or wind energy. More people in more cities are participating in curbside recycling programs, rinsing and sorting glass jars, plastic jugs, and tin cans, as well as sifting through piles of cardboard, paper, and newsprint.

What, then, is the net result of the strengthening of environmentalism? Most notably, there have been significant gains in energy and resource efficiency across the globe. Loggers and fishers are striving to maintain consistent yields. Processors and manufacturers are upgrading to conserve energy. Truckers and shippers are packing to reduce waste. Supermarkets and department stores are selling more with less shelf space. Consumers are turning off lights and unplugging appliances to lower electric bills. Recyclers are returning resources to the manufacturing cycle in increasing quantities and with increasing rates of efficiency. New product designs are making the recycling stage easier and cheaper. And waste management firms are doing a better job at treating, incinerating, and disposing of garbage.

This is creating a global economy able to produce more consumer goods with less energy and less waste per unit of output. At the same time, more people are purchasing "fair trade," "organic," and "sustainable" products; more are now recycling and conserving; and more are using products more efficiently. Indeed, it's possible, as Bjørn Lomborg did in his bestseller *The Skeptical Environmentalist*, to fill a book with hundreds of pages and thousands of footnotes of good environmental news. Still, such good news does not mean all is well with the earth as a whole.[7]

The Failures of Incremental Environmentalism

Stepping back to look at our global environment reveals a disturbing picture. The rising tide of consumption worldwide is swamping many of the gains from stricter environmental laws, higher environmental standards, and the creative energy of environmental activists and philanthropists. It's also swamping the emerging environmental markets and the lower per unit environmental impacts of manufacturing. Because *total* consumption never falls unless an economy plunges into a depression or a society implodes into civil war or anarchy, the resulting numbers are daunting. Billions of people are now consuming vast quantities of everything, and the totals keep rising. There are thousands of examples. Some, like the annual production of 5 trillion or so plastic bags, have so many different sources and uses it's not even really possible to keep track.

This relentless rise in consumption in a globalizing political economy of rising trade and investment is casting ever longer and deeper ecological shadows alongside—or sometimes *within*—the progress of global environmentalism. The solutions to problems posed by consumer goods almost always involve producing more of other goods, as in the case of cars: more car seats, seat belts, air bags, roads, traffic lights, and parking spaces. The solutions seldom involve producing *less* of a good—as with CFC refrigerators or leaded gasoline—and, even then, replacing that good with another almost always leads to consuming more overall.

Here, as the cases in this book show, the pursuit of profits and economic growth tends to supersede calls for precaution, even in situations of high uncertainty and risk. As the globalization of trade and investment extends the distances between producers and consumers, the resulting process of change tends to displace consequences along ever longer trade and corporate chains connecting distant regions, from Africa to Asia to the Antarctic. Effects spill into the future as well, sometimes taking generations to appear. This obscuring and displacing of environmental costs makes it harder for consumers to perceive—and thus care about—the cumulative impact of seemingly inconsequential personal choices on the global environment. Moreover, producer and consumer prices of many traded products, such as timber and beef, do not reflect the full environmental or social costs of harvesting, processing, producing, transporting, marketing, or disposal. As such, they reduce revenues available for environmental management, a particular problem for regions that rely on natural resource exports to generate economic growth. The resulting low prices for consumer goods made from these resources then stimulate wasteful consumption and overconsumption in importing countries—which helps explain why, for example, supersizing is so profitable for fast-food chains and obesity is increasing worldwide.

Over time, as sovereign states and multinational firms pursue their interests and cost-effective solutions, and as international financing props up economies with foreign debt, a disproportionate share of the environmental costs of such consumption tends to be shifted onto poor people and into ecosystems at risk—from the slums of India to the rainforests of Cambodia. Because such people and places tend to have less capacity to adapt to resulting changes, this further intensifies the consequences of consumption. As the history of leaded gasoline shows, a phasedown of a dangerous product in some jurisdictions can lead to its export with attendant environmental costs to other jurisdictions in distant lands.

The conclusion here is deeply troubling. Not only is environmentalism failing to produce sustainable patterns of global consumption, much of what policy makers in high-consuming economies are labeling as "environmental progress" is in reality little more than the wealthy world deflecting consequences and risks into ecosystems and onto people with less power—and thus less influence over global affairs.

This in part explains why support for more economic growth among the ruling elites remains rock-solid even with clear signs of an escalating global environmental crisis. Many assumptions buttress this. It's an individual's right to consume. It's a corporation's function to offer competitive choices. And it's the duty of a community to ensure the (increasingly material) well-being of its members, of a government to ensure steady economic growth and job opportunities for its citizens, and of an international lending institution like the World Bank to stimulate global economic growth. A near consensus exists on the best path forward to enhance human welfare and promote a more sustainable form of development: more free trade, higher per capita incomes (in real terms), independent multinational companies, responsible global financing, competitive markets, and sound scientific research. The widely held belief is that trade and investment in competitive markets ensure the efficient allocation and use of resources. Financial assistance is necessary here to stimulate growth in less-developed economies, which in turn helps to keep the global order stable. At the same time, scientific research—seen as objective and rational—ensures both that technological progress will occur, thus improving consumer choices, and that the potential dangers of introducing new products will be accurately assessed.

One effect of this near consensus—or what some call "ideology"—is to empower industry scientists, thus enhancing the capacity of companies to obfuscate, placate, and generate uncertainty about the need to act (as well as to bolster fears about the economic consequences of "unnecessary regulations"). They did just that after independent scientists began to investigate the environmental presence and consequences of lead in the late 1960s and of chlorofluorocarbons in the 1970s. Despite much progress since then, today such corporate tactics continue to delay, block, and even gut many environmental regulations.

The technologies of globalization—planes, phones, computers—allow critics of the global order to communicate, and sometimes, as with anti-globalization activists, even organize vocal protests with worldwide media coverage. On the surface, this would seem to enhance the power of critics to induce global change. Yet the process of globalization is embedded in the world economic order—in the production and trade chains of the biggest corporations and most powerful states. For this reason, the net effect of globalization is to reinforce and expand the global culture of capitalism rather than to empower voices critical of consumerism.

Those in power tend to dismiss or ignore critics who argue that the structures of global interactions—free trade, multinational corporations, the United Nations system—give rise to unequal consumption, overconsumption, and wasteful consumption. The policy and corporate worlds label such critics "unrealistic," "irresponsible," and "hypocritical." A few even label them "racist neocolonialists" for denying the poor of the developing world the *right* to consume. Thus calls to reverse economic globalization or to localize trade gain little traction even among nonprofit environmental groups. The world community is gravitating instead toward environmental solutions that fit into—and reinforce—the neoliberal economic order. Many of the buzz phrases of environmentalism embrace a corporate worldview: "business-NGO partnerships," "eco-efficiency," "corporate social responsibility," "voluntary compliance," "market mechanisms," "technology transfers." Even most nongovernmental organizations, in a compromise for relevancy and funding, are now focusing on small, achievable steps, on, for example, partnering with firms and states to improve the management of a particular ecosystem.

The partnering of some NGOs with governments and firms doesn't mean activists are no longer challenging the established order. If anything, they're doing so more than ever before, in part, because the Internet now provides them a global forum that is both cheap and easy to use. This can contribute to a greater diversity of input, including input from people far removed from the centers of power. Yet it can also produce a cacophony of critical voices over some issues—with the paradoxical result that none are "heard." And the globalization of markets can make it hard for these groups to influence consumers, especially as more and more public relations wings of firms and governments counter environmental criticisms in the mainstream media. This is the case with the current campaign to end Canadian sealing. Although the Western media will still cover a Paul McCartney landing to protest the hunt, his message and those of other protestors have little influence on the major consuming markets of Russia or China. Even the North American campaign to boycott the Red Lobster restaurant chain has garnered little support—in part, because the Canadian government has become more adept at advertising its "side" of the "truth."

To reiterate, small, achievable changes *are* helping to mitigate particular environmental impacts of particular forms of consumption. Yet changes to mitigate the impact of *global* consumption on the earth's environment remain too slow and incremental to avoid irreparable damage. The evidence of ongoing—and increasing—harm to people, forests, deserts, freshwater, oceans, and the climate is overwhelming. The conclusion here is inescapable: despite much progress and prosperity over the last half century, if the world hopes to avoid an even greater crisis by the middle of this century, it must transform and accelerate the processes of environmentalism. This raises a final—and most difficult—question: How?

A Brighter World Order of Balanced Consumption

Transforming environmentalism to control the shadows of consumption will take years of consultations and negotiations. The following musings are therefore intended simply as a way to begin a conversation. Working toward more "balanced consumption," I submit, has the potential to mitigate many, if not most, of the damaging ecological effects both of individual consumption and of the corporate, trade, and financing structures producing consumer goods.

Any lasting progress toward more sustainable global patterns of consumption will require a mix of policies and incentives. Wealthy consumers can assist by pursuing more personal balance between needs and indulgence—by practicing, in the language of Thomas Princen, Michael Maniates, and Ken Conca, more "cautious consuming" guided by values like "thrift," "frugality," "simplicity," and "self-reliance."[1] All consumers can work toward balance by reusing, conserving, and, to a lesser extent, recycling, more consumer goods. But, as the analysis in this book shows, consumers alone cannot significantly diminish the ecological shadows of consumption, much less eliminate them. The world must reorient the structures and processes that guide consumption to create more balance across and within societies, reducing the inequalities of consumption, and allowing living standards to improve for the poor without doing grave harm to the global environment. Any transformative change must also address the imbalances between the inputs into growing economies and the sustainability of ecosystems.

Achieving more balance will demand measures to rein in some of the damaging environmental consequences of multinational corporations, trade, and financing. I see the following as a necessary beginning. Multinational corporations will need to eliminate double standards for domestic and foreign operations. Manufacturers will need to take a more precautionary approach to replacing one technology with another. The prices of traded goods will need to better reflect ecological and social costs. States will need to ensure that trade and trade agreements do not lower environmental standards. And more financial aid will need to go toward compensating poor people living in degraded ecosystems. International laws on the environment will need to be so well drafted and enforced that firms and states cannot meet obligations simply by casting ecological shadows elsewhere. Governments will also need to create incentives to reduce overconsumption and wasteful consumption, aiming to protect the earth itself and not, as occurs now, merely their respective countries. Activists will need to campaign to expose these shadows as well, so that everyone begins to see national initiatives that deflect harm elsewhere as *causes* of the escalating global environmental crisis, not as "solutions" or "sustainable development."

To address this tall order, let's first take a closer look at how consumers can help.

More Balanced Consumers

Billions of consumers are now reusing or recycling items like bottles, cans, and newspapers. The world needs responsible consumers, and most efforts to reuse or recycle goods promote sustainability. Not all, though, involve straightforward gains. Thus curbside recycling in many cities uses trucks to collect discarded goods and factories to sort and clean them, requiring both money and energy while also producing pollution. On occasion, for lack of buyers, authorities end up dumping recycled waste into landfills. Many consumers are unaware of what happens after dropping a can or bottle in a recycling bin; for others, this act alone seems to alleviate their worries about the consequences of drinking from a Styrofoam cup or driving a 5,000-pound SUV.

Still, personal efforts like recycling bring many benefits even when the outcomes don't match those promised. This book contains a stream of examples of people choosing to reduce waste and environmental damage. More people are choosing careers as "environmentalists"—from activists opposing the seal hunt, to policy makers in environmental agencies, to environmental analysts in the World Bank. Many others are making smaller efforts, buying more energy-efficient appliances and trading in gas-guzzlers for hybrid cars. More people are turning off unnecessary lights as well as turning down heaters and air conditioners. More people are eating organic beef and drinking fair-trade coffee. More people are buying eco-certified timber or seafood. More people are boycotting real fur coats. And just about everyone is now fueling cars with unleaded gasoline. The list goes on and on.

Some consumers, then, are altering *some* practices. Whether motivated by environmental concerns or by self-interest (such as saving on energy bills), these consumers are lessening the impact of personal consumption of *particular* goods on the global environment. This is a positive trend, and this book contains convincing evidence that it's gaining in strength. Still, such efforts alone do not have the power to transform *global* patterns of consumption casting ecological shadows. As a global constituency, consumers are too diverse—and themselves too changeable—to induce enough lasting change. Moreover, this diversity and changeability is increasing as the current wave of economic globalization shifts markets at even faster rates across even more cultures.

Thus, as this book has shown again and again, the crosscutting choices of consumers even within one state tend to increase environmental impacts overall. Even as sales of hybrid cars in the United States are rising, for example, sales of sport utility vehicles and other light trucks are climbing toward half of all vehicle purchases. Such crosscutting tendencies are magnified as markets go global. Consumption of Canadian seal furs is a good example. With each passing year, as chapters 20 and 21 have charted, the activist campaign to convince consumers that Canada's seal hunt is immoral and unsustainable has been gaining supporters. After getting the European Union to ban the import of whitecoat pups 6–12 days old, it even managed to bring an end to the commercial hunt for harp seals in the 1980s. To this day, few Europeans will buy (or wear) a fur from a harp seal (even from the older pups). Yet, because demand for seal furs in other cultures, notably, those of Russia and China, is high and growing higher, the seal hunt is now even larger than before the campaign began in the 1960s.

The "sustainable" meat industry is another example of the power of globalizing markets to swamp the benefits of changing local purchasing patterns. Consumption of organic beef and chicken is increasing in North America and Europe; yet pound-for-pound the consumption of meat from industrial farms is rising there even faster. Worldwide, consumption of industrial meat has increased more than fivefold since 1950, and per capita consumption has more than doubled.

Educating consumers and expanding green markets within particular cultures *can* reduce individual impacts and allow economies to produce more goods with fewer inputs. Eliminating ecological shadows, however, will require far more than educating citizens and expanding environmental markets. It will require efforts to prevent environmentally harmful products, with their attendant ecological costs, from being shifted into "emerging" markets as sales of such products in "educated" and "green" markets shrink. And it will require tough measures to transform how multinational corporations, trade, international financing, and state policies currently distribute the costs and benefits of globalized consumption.

Balancing Corporations

Although humans must consume to survive, meeting basic needs accounts for only a fraction of the environmental damage from rising consumption. Much of this damage arises when corporate advertisers, playing with words, facts, and perceptions, induce consumers to go far beyond basic needs, to supersize their desires, to recognize new "needs." What is safe? Necessary? What is healthy? Desirable? To increase their profits and market shares, corporations have long answered these questions in ways that confuse—and at times deceive—consumers. Do we really need to kill every germ to protect our children? The makers of the disinfectant Lysol—who claim it kills 99.9 percent of harmful germs—say we do: "Life demands Lysol. That's a fact." It's easy to laugh at such blatant hype. Yet corporate indifference to truth in advertising—or truth in research—rarely produces a laughable outcome.

The tobacco industry is a notorious example. Today, no objective physician would deny the deadly effects of smoking. The World Health Organization estimates that illnesses related to tobacco—notably cancer, strokes, and heart attacks—killed around 100 million people in the twentieth century alone. Still, even with millions dying, cigarette companies continue to manipulate and conceal research, delay and block regulations, work the courts as part of standard business, and target ads at young people. Governments in many wealthy countries strive to "protect" consumers from these firms both by educating them about the perils of smoking and by mandating nonsmoking areas. As a result, tobacco sales in these countries are declining. At the same time, however, *total* sales continue to climb worldwide as more and more people take up smoking in developing countries. The health consequences are appalling: smoking kills about 5 million people every year—a number the World Health Organization predicts will double by 2020.

There are also many instances of corporations exporting harm to offset declining sales at home. In some cases, national policies put in place to protect citizens and local environments in wealthier countries create incentives for companies to export ecological costs to poorer places. The resulting consequences can be as appalling as those of tobacco. International environmental agreements can reduce, mitigate, and even prevent some of the consequences of these corporate double standards, as the agreements to protect the ozone layer have clearly shown. Corporate technology transfers and international aid can also accelerate processes within poorer regions to deal with the shadow effects arising from different corporate standards. In rare instances, corporate codes of conduct can create opportunities for multinational corporations to hold affiliated suppliers to higher standards than local laws require. Change can even occur swiftly when corporate interests in expanding new markets trump those profiting from old ones—as happened after the 2002 World Summit on Sustainable Development, when it took just four years to replace leaded with unleaded gasoline across sub-Saharan Africa.

But such efforts alone are often too slow—and too late—for the people and ecosystems most at risk. Governments at all levels need to put in place much tougher measures and disincentives to prevent the unequal practices of multinational corporations from displacing ecological costs. NGOs can assist here by campaigning to reveal these costs and to highlight double standards—while consumers can help by boycotting corporations that continue to profit from displacing them. A binding international code of conduct would greatly increase the capacity of states to control the harmful behavior of multinational investors and corporations.

Mitigating the tendency of exporters and investors to deflect the ecological costs of consumer goods will require corporations to respond to new environmental rules and norms proactively—with *precaution*. All too often in the past, they have responded to environmental measures by denying, delaying, then gradually *replacing* the "problem" products with "improved" ones (often doing so in wealthier markets first). And, as the histories of leaded gasoline and CFC refrigerators show, a common result is to replace one ecological stressor with another: lead with benzene, chlorofluorocarbons (CFCs) with hydrofluorocarbons (HFCs). At the same time, in pursuit of greater profits and larger markets, they have kept developing and selling new consumer goods. Many of these goods continue to bring consumers convenience or pleasure. Some, like air bags or smoke alarms, make life safer, and a few, like computers, transform people's lives, altering everything from relationships to global commerce.

The value of innovative corporations extends beyond serving consumers better: they can also help economies and societies prosper.

Nevertheless, introducing new goods runs the risk of replacing present harms to people and ecosystems with future ones. How safe are the wonder chemicals in nonstick pots and pans? Or in fire-resistant rugs, mattresses, and pillows? No one really knows for sure. Only a handful of the tens of thousands of chemicals in consumer goods have undergone rigorous testing for harmful side effects. Could such effects partly explain the rising rates of cancer in countries like the United States? Preliminary testing on just a few thousand has shown that hundreds of these chemicals can cause tumors in laboratory animals. What will happen when these chemicals accumulate, combine, and age? Again, no one knows for sure.

What we *do* know is that corporations routinely introduce new products with little understanding of the environmental consequences. No doubt the benefits of many will outweigh any future costs, but some will prove harmful, even deadly to people and ecosystems. It's naive to assume they won't. Who in 1928—or 1938 or 1948 or 1958—could have foreseen the global aftermath of Thomas Midgley's award-winning discovery of wonder refrigerants—stable, nontoxic, nonflammable chlorofluorocarbons? Who could have foreseen that CFCs would drift into the upper atmosphere and, generations later, deplete the ozone layer? Indeed, no one even *conceived* of such a possibility, at least not publicly, before the 1970s.

Still, firms and governments, by taking a precautionary approach, could do far more to reduce the risks of introducing or modifying consumer goods. What would this involve? The core idea, explains Kerry Whiteside, professor of government at Franklin & Marshall College, is for governments and firms to follow a "strategy of anticipatory preventative action." Such a strategy demands "*better science* and more self-conscious *political judgments*." Better science and politics, in turn, means "doing science *differently*, with more dialogue between practitioners in disparate disciplines and more transparency in relation to the nonscientific community." It also requires far more humility because "progress consists in recognizing our *inability* to master the world."[2]

Taking a precautionary approach also requires a critical change in thinking—accepting that the burden of proof that a new chemical, organism, or device is not harmful must rest with those proposing to introduce it into an ecosystem. Here it's necessary to "resist the temptation to believe that every technological risk is worth taking or that we will be able to repair whatever damage we do to our surroundings." It's necessary as well to ensure that those who harm others compensate them, even if it takes many generations for the harm to appear. Developing a precautionary approach to protecting the global environment further demands institutions and policies that embrace "not only fellow citizens in one's own nation-state but also people across the globe and their successor generations."[3]

Such changes will not come about easily. The U.S. government rejects precaution as a basis for most of its policies and decisions—and so do most other governments—arguing it would stifle innovation and slow economic growth. Still, over the last few decades, precautionary thinking has been gaining traction in a few places, although slowly, unevenly, and under various definitions. It first emerged in environmental legislation in Germany in the 1970s, then spread through Europe in the 1980s and 1990s. Today, many European governments as well as some international treaties and documents refer to it; France even amended its constitution in 2005 with an environmental charter that includes a precautionary principle in Article 5.

Nevertheless, the *global* effects of these changes in Europe remain weak. A key reason is the reaction of multinational corporations to a call for more precaution. Almost all now display a public profile of responsible investment, with thousands of businesses in over 100 countries currently members of the UN Global Compact, which states in principle 7: "Businesses should support a precautionary approach to environmental challenges."[4] This may sound promising, but, thus far, corporate practices haven't come close to living up to that promise. No multinational corporation takes a true precautionary approach to environmental management. Many still invest after only perfunctory assessments of environmental impacts; many continue to mine natural resources under unrealistic models for sustainable yields; many still make and transport goods without accounting for ecological costs; and most continue to introduce new products after only limited research that tends to downplay any risks. Here the global trading system, spreading consumer goods across countries with ever-increasing speed, can cause a seemingly small corporate error to grow into a force powerful enough to damage the earth's capacity to sustain life.

Balancing Trade

No simple solutions exist to mitigate the shadow effects of trade. States can neither abandon nor severely restrict international trade—within current economic structures, too many societal benefits flow from exchanging goods and services across borders. Yet measures are still necessary to protect vulnerable people and ecosystems. International agreements to restrict, if not ban, trade in endangered species and the export of hazardous waste can help. Eco-labels and organic markets can also shift demand toward products from more sustainable sources or with smaller environmental impacts over a life cycle of use (such as more energy-efficient appliances). Government policies can help as well to ensure that consumer prices reflect more of the social and environmental costs of natural resources, manufacturing processes, and transportation. But restricting trade with import tariffs or export bans can also distort prices and incentives, causing environments to degrade at even faster rates and protecting inefficient producers, which, in turn, can reduce government revenue—collected through levies or taxes—for environmental management. The history of logging in the tropics is full of such examples; government policies to control the timber trade and protect domestic processors partly explain rising deforestation rates in some tropical countries.

States and international institutions like the World Trade Organization need to guide global trade with anticipatory strategies to prevent ecological shadows, as do corporations when investing in new places or marketing new goods. This requires stronger measures to ensure that trade and trade agreements don't serve to lower environmental standards. It also requires greater efforts to include the ecological and social costs in the prices of consumer goods, both to raise revenues for local communities and to reduce wasteful consumption and overconsumption. Currently, prices of many, if not most, traded goods do not adequately account for the ecological and social costs. Take the price of a typical piece of tropical plywood: although it generally reflects the corporate costs of logging, processing, transporting, retailing, fees, and taxes, it does *not* reflect either the ecological costs to the rainforest and local animals—and to the global climate over the long term—or the social costs to the local people. This is equally true for many other traded products, such as beef from the Amazon.

On the other hand, believing that the answer lies in localizing trade or banning international trade outright can cause as much harm as believing "free" trade will solve all problems. Although specific, targeted international trade bans are certainly necessary to protect species nearing extinction, state or regional bans can backfire, causing new markets to emerge with even fewer environmental controls. Moreover, trade restrictions can protect inefficient firms and stimulate wasteful consumption, as was the case for plywood exports from Indonesia during the 1980s and 1990s. Undercutting the economic or social value of an ecosystem can produce compelling incentives to convert it into a different ecosystem with higher value—such as burning down a logged rainforest to develop a soybean plantation or cattle ranch.

International institutions need to hold states more accountable for the *global* environmental effects of restricting trade for local reasons. In many instances, global costs outweigh local benefits. China's decision to ban logging in the late 1990s to control flooding, for example, caused timber imports to soar when China's booming construction industry sought new sources of cheap lumber. The result was predictable. Exporters like Indonesia saw deforestation rates soar: from 2000 to 2005, the outer islands experienced the world's highest rates of forest loss. Most of the short-term ecological costs—flooding, soil erosion, forest fires—are confined to relatively remote places like Kalimantan or Sulawesi. But the ecological shadow from China's consumption of tropical timber will one day sweep back over China: tropical deforestation adds up to one-fifth of the human-induced carbon emissions now causing climate change.

Although international laws and national policies are preventing some of the shadow effects of trade, overall, as organizations like the World Bank and World Trade Organization pressure developing countries to open markets and to welcome foreign investment, it's becoming easier, not harder, to displace the ecological costs of producing, using, and disposing of consumer goods. Far more needs to be done to ensure that when states and firms engage in international trade, they don't dump used goods, like cars or computers, into poor places without the capacity either to recycle them or to enforce reasonable environmental standards. During phasedowns of dangerous products, more also needs to be done to prevent states and corporations from using trade to compensate for falling sales at home by expanding sales in markets with weaker environmental rules.

At the same time, international institutions and national governments need to do more to monitor and guide the process of replacing consumer goods through trade. Although trade can accelerate changeovers in developing countries, it can also end up replacing a small problem with an even bigger one when the *total* impact of newer products outpaces the gains from replacing the older ones.

The auto industry is a good example of the potential dangers of replacing an older product with more and more of a "safer" newer one.

Statistics show that, thanks to better technologies and controls—seat belts, infant seats, shatterproof glass, antilock brakes, paved roads, traffic lights, speed traps, drunk-driver checkpoints—riding in a car today is becoming less and less dangerous in wealthy countries. The list of advances is getting longer with each passing year. Yet the simultaneous increase in the number of automobiles—traveling farther over every "pavable" square inch of the planet—means the overall impact on human safety is now a global crisis. Traffic collisions from the more than 800 million passenger cars and commercial vehicles, which every year injure between 20 and 50 million people and kill more than 1 million, are now the biggest cause of violent injury and death globally. And these numbers are continuing to rise as the number of "improved" cars heads toward the 1 billion mark. Indeed, experts now predict that annual traffic deaths will reach 2 million by 2020.

The history of tailpipe emissions is similar. Thanks to technologies like catalytic converters and hybrid engines and government regulations like California's Smog Check Program, many vehicles now emit less greenhouse gases per mile traveled. The air quality in many California cities is better as a result of such advances. Thus far, however, the sum of all of these advances has produced only incremental decreases in environmental impacts—nothing approaching the dramatic decreases that could be achieved by replacing gasoline engines with, say, hydrogen fuel cells, which emit only water and heat. And, in 2005, the American Lung Association still ranked the Long Beach and Riverside traffic corridor of Los Angeles as having the most polluted air of any city in the United States. Rising pollution from traffic in cities like Shanghai, Mexico City, and Delhi over the last few decades, moreover, counters any incremental gains in London or Vancouver or Los Angeles. Rising numbers of gas-guzzling SUVs and increasing numbers of used car exports to developing countries further strain the global environment. The global trend in tailpipe emissions parallels that for traffic safety: new technologies and regulations are *reducing* the tailpipe emissions *per car* in wealthier countries, yet automobile emissions *worldwide*—and especially in poorer countries—are *rising*, with particularly harmful effects on poorer people.

It's essential, then, to guide the replacement of products with greater care and precaution. The global effort to replace CFC refrigerators shows the potential for trade—coupled with sound international agreements, cooperative multinational companies, reasonable international financing, and consistent state policies—to accelerate environmentally friendly change worldwide. This effort also shows how, as part of the process of replacing consumer goods, further environmental gains are possible (in this instance, many of the new refrigerators use less energy).

Many other opportunities exist for similar processes of change to occur. Take one seemingly minor example, the incandescent lightbulb. Though a great advance for the late nineteenth century, incandescent lightbulbs are still highly inefficient: less than 10 percent of the power they use produces visible light. By contrast, compact fluorescent lightbulbs use as little as one-quarter of the power to produce the same amount of light—and they last up to ten times longer. According to the International Energy Agency, switching to fluorescent lights could decrease worldwide use of electricity by 18 percent. Progress toward replacing incandescent lightbulbs has been slow, however, in part, because they're cheaper than fluorescent ones. Incandescent lightbulbs still account for 67 percent of lightbulb sales—though only 4 percent of light output—worldwide.[5] On the other hand, there are some signs of greater progress. In 2007, the Australian government announced it would phase out incandescent lightbulbs, with the goal of cutting the country's greenhouse gas emissions by 4 million metric tons by 2012—and household power bills by up to 66 percent. New Zealand and Belgium are considering similar phaseouts, as are California and New Jersey.

Such changes, even small ones like replacing inefficient lightbulbs, will not be effective without financing (for example, to ensure safe disposal of the mercury in burned-out fluorescent lightbulbs). More far-reaching ones, such as conserving energy and resources with green architecture, will require far greater investments. Governments, firms, and consumers in countries like Australia, New Zealand, Belgium, and the United States will need to help pay the transitional costs of these changes and do far more to balance the financial flows between the First and Third Worlds.

Balancing Financial Flows

Bilateral and multilateral aid helps environmental management in developing countries in many ways. It funds state programs to meet environmental commitments under international treaties, and it supports corporate efforts to integrate environmental technologies into harvesting and manufacturing processes. It finances environmental research and education and provides technical assistance for government environmental agencies. It also funds nonprofit environmental organizations working within communities as well as those partnering with firms and state agencies. And it helps consumers replace harmful goods with less harmful ones.

International aid can also strengthen the capacity of weak states to block ecological shadows, enhancing, for example, the ability to monitor and enforce environmental laws. It can provide essential support for collaborative efforts to protect ecosystems beyond the control of one sovereign state, whether the high seas, the stratosphere, or Antarctica, from the ecological costs of consumption. It can enhance, too, efforts to follow a precautionary principle for environmental decision-making within developing countries. It can accelerate efforts to replace consumer products harming the global environment with environmentally friendly ones by enhancing the capacity of governments, firms, and consumers in countries like China and India to meet international environmental commitments, as in the case of CFCs.

At the same time, however, international aid can leave people and ecosystems even more prone to absorbing the ecological costs of rising global consumption. Grants and technical assistance tied to trade and investment interests can do more to guarantee cheap goods for consumers within donor states than better living conditions within developing countries. Loans as "aid" are an even bigger problem. Interest rates and terms on these loans are generally better than a commercial bank would offer. Still, for many decades now, most borrowers have been struggling to repay the interest on, let alone the principal of, these loans, and, with economies (and currencies) in boom-and-bust cycles, almost all have ended up borrowing even more to survive. One consequence is an escalating external debt across the developing world—now more than 30 times higher than in 1970, with "aid" recipients paying donors over $100 billion just in interest.

Over the last half century, many of these states have been ramping up natural resource exports to earn enough foreign exchange to service this debt. At the same time, organizations like the World Bank and International Monetary Fund (IMF), in an effort to stimulate foreign exchange earnings, have been requiring governments to remove controls on trade and capital flows as a lending condition. The resulting "structural adjustments" have shifted priorities even more toward exporting natural resources and low-end manufacturing—creating economies of tin, timber, and T-shirts. These adjustments have also left many developing countries more open to firms dumping waste and used goods from high-consuming countries. Some economies do stabilize—and even grow—following these adjustments. But, too often, they leave themselves open to the ecological shadows of high-consuming "donor" states, turning rice paddies into shrimp ponds and towns into smog factories.

International donors need to distribute grants and technical assistance that serve the interests of people and ecosystems in developing states more than the financial interests at home. And they need to transfer far more funds into developing states to assist with efforts to protect the *global* environment. Although the Global Environment Facility, with over $7 billion in grants from 1991 to 2008, is a reasonable beginning, high-consuming states should be transferring hundreds, not tens, of billions, and not out of goodwill, but to mitigate and *compensate* for the shadows of their rising consumption. Debt relief needs to be far reaching, able to break the cycle of poor countries servicing rising debts by exporting more and more goods to consumers in donor states, goods kept cheap and plentiful by exploiting people and drawing down the globe's natural capital.

Navigating the Future

Sweeping reforms to the world order, then, are necessary to accelerate and deepen efforts to balance consumption. The globalization of environmentalism is improving management on *some* measures, significantly decreasing the per unit impacts of *some* consumer goods for *some* consumers. But it's failing to prevent the globalization of investment, trade, and financing—powered by multinational corporations and strong states—from displacing a disproportionate share of the ecological costs of rising consumption into the most fragile ecosystems, onto the poorest people, and into distant times. Concentrating ecological impacts on the most vulnerable is not only unjust for billions of people; it is also far more likely to tip societies and ecosystems into uncontrollable decline and collapse.

Any chance of transforming the environmental consequences of global consumption will require far greater efforts than a book scratching at the surface of how and why ecological shadows form, shift, and fade. That said, I believe a reasonable starting point is to pursue more balanced consumption—personal and structural—so that less of the costs and more of the benefits of producing, using, and disposing of consumer goods are shifted to the world's poorest people and most vulnerable ecosystems. We must all become more responsible consumers. Our environmental standards must be higher, our technologies better and less wasteful, our eco-markets bigger, our progress toward environmental targets faster. But all this, by itself, will not be enough. To reform the structures causing deep imbalances in consumption, international agreements and organizations will need to do a far better job in guiding economic globalization and in restraining the self-interest of sovereign states and multinational corporations. To achieve more equitable patterns of global consumption, governments and producers alike must hew to an exacting precautionary principle, under which multinational corporations follow consistent standards across jurisdictions, the prices of traded goods reflect more of the environmental and social costs, and more international aid compensates for the effects of consuming so much of the natural capital of the developing world. Only then can we begin to navigate toward a brighter future.

Notes

Chapter 1

1. This calculation uses the 2007 Hummer H3 weight of 2.2 metric tons (4,800 pounds). The estimate of global carbon dioxide emitted by the "consumption and flaring of fossil fuels" is from the U.S. Energy Information Administration, at http://www.eia.doe.gov/.

2. From the back cover of the storybook *"Slowly, Slowly, Slowly," Said the Sloth* (Carle 2002).

3. More precisely, total world GDP, in constant 1995 dollars, went from just under $8 trillion in 1960 to $35 trillion in 2002 as the world's population grew from 3 billion to 6.2 billion—a per capita increase of $3,000, from $2,600 to $5,600. See World Bank, World Development Indicators Online, at http://www.worldbank.org/.

4. UNFPA 2004, 8; United Nations Secretariat 2001.

5. World Bank 2005, 236.

6. Of the handful of books that address the environmental impacts of consumption, one of the best is *Confronting Consumption* (Princen, Maniates, and Conca 2002), a collection of articles that examine the "ecological political economy" of consuming, analyzing in particular "the social and political, especially the organized exercise of influence and power to skew benefits to some and harm to others" (p. ix). In particular, see Princen 2002a, 2002b; Maniates 2002a, 2002b; Manno 2002; Conca 2002; Clapp 2002; Tucker 2002; Helleiner 2002. For other helpful books and articles, see Redclift 1996; Goodwin, Ackerman, and Kiron 1997 (a collection of classics); Westra and Werhane 1998; Crocker and Linden 1998; Lichtenberg 1998; Luban 1998; Rosenblatt 1999; Schor and Holt 2000; Woollard and Ostry 2000; Cohen and Murphy 2001; Myers and Kent 2004; Southerton, Chappells, and Van Vliet 2004; Ehrlich and Ehrlich 2004; Schor 2004, 2005; Clapp and Dauvergne 2005; O'Rourke 2005; Cooper 2005; Wapner and Willoughby 2005; Fuchs and Lorek 2005; Princen 2005; Greenberg 2006; Paterson 2007.

7. As far as I know, the term "ecological shadow" first appeared in MacNeill, Winsemius, and Yakushiji 1991, 58–59. I first used the concept in *Shadows in*

the Forest (Dauvergne 1997), where my purpose was "to evaluate the environmental impact of one country's economy on resource management in another country or area" (p. 2). The 1997 book focuses on the impact of Japan on the rainforests of Southeast Asia, revealing how Japanese traders and financiers were interacting with local politics to accelerate deforestation. The current book extends the concept of ecological shadows to capture the more complex features of interlacing patterns of global consumption.

8. Sociologist Roland Robertson (1992, p. 6) is commonly credited with having first described the globalizing world as "a single place." The literature on globalization is too vast to list in its entirety. See, for example, Hirst and Thompson 1999; Robertson and White 2002; Giddens 2002; Bhagwati 2004; Wolf 2004; Friedman 2005; Scholte 2005.

9. World Commission on Environment and Development 1987, 43.

10. Garcia-Johnson 2000.

11. World Bank 2006. See also World Bank, World Development Indicators Online, at http://www.worldbank.org/. The estimate of the foreign direct investment flows to developing countries in 2006 is from UNCTAD 2007.

12. See World Bank 2006 for details. See also World Bank, World Development Indicators Online, at http://www.worldbank.org/.

13. See Friedman 2005 for an analysis of what he calls the "flattening effects of globalization."

14. See McKendrick, Brewer, and Plumb 1982; Tucker and Richards 1983; Tedlow 1990; Tucker 2002.

15. World Bank 2006; UNCTAD 2002, xv, 272; UNCTAD 2001, 1, 9. See also World Bank, World Development Indicators Online, at http://www.workbank.org/.

16. Quotation is from Japanese Forestry Agency 1993, 2. See also Dauvergne 1997, 2001.

17. See Porter, G. 1999.

18. One of the best-known critics of the environmental effects of trade is Professor Herman Daly at the University of Maryland. See, for example, Daly 1993, 1996, 2005.

19. World Bank, World Development Indicators Online, cited in Clapp and Dauvergne 2005, 193–195.

Chapter 2

1. Because reliable local data are often hard to obtain, global environmental statistics are all rough estimates. To ensure some balance, I've drawn the statistics in this book from a range of sources. These include United Nations sources, in particular, the UN Environment Programme (http://www.unep.org), the UN Development Programme (http://www.undp.org), the Food and Agriculture Organization (http://www.fao.org), and the World Health Organization

(http://www.who.org). Besides those listed in the references, sources for double-checking and counterchecking include the World Bank (http://www.worldbank.org), the World Resources Institute (http://www.wri.org), and the World Wildlife Fund / World Wide Fund for Nature (http://www.panda.org).

2. FAO 2005a, 21.

3. Dalton 2005, 1056.

4. The Canadian Committee on the Status of Endangered Wildlife declared the northern cod endangered in 2003.

5. See Worm et al. 2006, 787–790.

6. See Myers and Worm 2003, 280–283.

7. Worldwatch Institute 2003; quotation is from the summary of the State of the World 2003, 9 January 2003, at http://www.worldwatch.org/. The estimated rate species are going extinct is from World Resources Institute 2005, 36.

8. Davis and Webster 2002, 25.

9. The U.S. statistics are from Wenz 2001, 5–6; the global ones from Green et al. 2005, 550.

10. On the role of the global political economy in climate change and altering various ecosystems, see, for example, Rowlands 2000; Litfin 2000; Newell 2000; Skjæreth and Skodvin 2001; Cass 2005; Bäckstrand and Lövbrand 2006; Victor 2006; Depledge 2007; Bulkeley and Moser 2007.

11. Stern 2006, ii.

12. FAO 2005b; see also IPCC 2001, chapter 11.

13. Canadell et al. 2007, 18866.

14. The United Nations Intergovernmental Panel on Climate Change (IPCC), which tracks and assesses research on climate change, asserted in a synthesis of the IPCC Fourth Assessment Report released in November 2007 that the evidence of "warming of the climate system is unequivocal." See "Summary for Policymakers," on the IPCC Web site, http://www.ipcc.ch/.

15. The rankings of warmest years rely on annual average temperature data from the National Aeronautics and Space Administration (NASA) for the earth's surface. See http://www.nasa.gov/. The average temperatures for 2005 and 1998 are close, but the 2005 highs arose without the warming effect of El Niño.

16. See the IPCC Web site, http://www.ipcc.ch/. Drew Shindell, as quoted in Zabarenko 2006.

17. As I use the term, "tipping point" is the point at which the process of steady decline in an ecosystem begins to accelerate uncontrollably toward a system crash. For a popular account of the social theory of tipping points, see Gladwell 2000; for a comprehensive environmental analysis of why some societies have collapsed and some haven't, see Diamond 2005.

18. Walter, Smith, and Chapin III 2007, 1657.

19. Thomas et al. 2004, 145–148; Pounds and Puschendorf 2004, 108. See Dressler and Parson 2006 for an overview of climate change.

20. Food accounts for about 95 percent of human exposure to persistent organic pollutants such as PCBs and dioxins (Betts 2004, 387A). The growing ecological impact of persistent organic pollutants also reveals the dangers of adding chemicals into local environments. See, for example, Clapp 2001; Selin and Eckley 2003; Downie and Fenge 2003; Stevenson 2005; Cone 2005; Maguire and Hardy 2006.

21. See Raloff 2003; Betts 2004; Hites et al. 2004; Bergeson 2005; Schecter et al. 2006.

22. Ascherio et al. 2006, 197–203.

23. Environment reporter Martin Mittelstaedt (2006a, A1, A8) summarizes these studies. Mittelstaedt's articles in the Canadian *Globe and Mail* on the environmental impacts of chemicals provide nice snapshots of recent research. See, in particular, his articles from 2006 and 2007. For a review more critical of research claiming that current amounts of PFOA in the environment are a threat to human health, see Weiser 2005.

24. Richard Wiles, David Boothe, and the EPA, as quoted in Mittelstaedt 2006a, A8. The Environmental Working Group is based in Washington, D.C. See their Web site, http://www.ewg.org/. See also Renner 2004, 1887; Stokstad 2006, 26–27.

25. The phrase "science-based approaches" is taken from the DuPont company response to those wanting to restrict chemicals like PFOA, as quoted in "Chemicals' Makers Criticize Ban," *Globe and Mail* (Canada), 20 June 2006, A8. Steven Hentges and Frederick vom Saal, as quoted in Mittelstaedt 2006b, A3.

26. See Davis and Webster 2002, 15–16.

27. A review of scientific publications on breast cancer and environmental pollutants, for example, found more than 200 chemicals causing mammary tumors in animal tests. Of these, 73 are or have been in consumer goods or food chains, 10 are food additives, 35 are in the air, and 25 are common in workplaces. See Rudel et al. 2007, 2635–2667.

28. Ana Soto, as quoted in Mittelstaedt 2006c, A8.

29. On coffee, see Wild 2004; Talbot 2004; Bacon et al. 2008; on bananas, see Striffler 2002; Striffler and Moberg 2003; Bucheli 2005; Soluri 2006; on sugar, see Schmitz 2002; Gudoshnikov, Jolly, and Spence 2004; on tea, see Moxham 2003; MacFarlane and MacFarlane 2004; on whales, see Stoett 1997; Heazle 2006; on elephants, see Pearce 1990; on tigers, see Meacham 1997; on pigs, see Jones 2003; on fisheries, see Bhattacharya 2002; Clover 2004; on forestry, see Dauvergne 1997, 2001; on mining, see Jackson and Banks 2003; on biodiversity, see Steinberg 2001; Mushita and Thompson 2002; on pesticides, see Hough 1998; Hond, Groenewegen, and van Straalen 2003; Pretty 2005; on coal, see Freese 2003; on hazardous waste, see O'Neill 2000; Clapp 2001; Pellow 2007; on persistent organic pollutants, see Downie and Fenge 2003; Johansen 2003.

Chapter 3

1. See "Inquests," *Times* (London), 21 August 1896, 6, col. B; "Inquest," *Times* (London), 26 August 1896, 4, col. F; see also Hamer 1996.

2. Based on data in WHO and World Bank 2004, 3, 172.

3. "Perpetual Motion," *Economist*, 4 September 2004, 4; U.S. Department of Energy, "Future U.S. Highway Energy Use: A Fifty-Year Perspective (Draft)," May 2001, as cited in McAuley 2003, 5415.

4. Gross et al. 1996, 83; McShane 1994, 135.

5. Henry Ford, as quoted in Gross et al. 1996, 84; see also p. 81.

6. Porter, R. 1999, 1; Jackson 1985, 161; Foster 1981, 58.

7. Foster 1981, 58; National Automobile Chamber of Commerce 1931, 26.

8. AMA 1935, 84; 1936, 84; 1938, 84; 1946–47, 30. The number of vehicles fell during World War II to, for example, about 31 million in 1944 (AMA 1944–45, 48).

9. Porter, R. 1999, 2.

10. AMA 1951, 28; MVMA 1978, 31.

11. Based on data in MVMA 1972, 28–29.

12. MVMA 1977, 28; 1982, 32; 1987, 36; AAMA 1993, 41. The advertising and GNP figures are from Freund and Martin 1993, 135.

The automobile industry as a whole constitutes an even larger percentage of the economies of many other developed states, accounting for around 10 percent of the GDP of the First World. "Perpetual Motion," *Economist*, 4 September 2004, 4.

13. AAMA 1997, 46; Ward's Communications 2002, 51.

14. Sheehan 2001, 60.

15. Statistics in this paragraph are drawn from Frumkin 2002, 201; Worldwatch Institute 2004a, 6; Worldwatch Institute 2004b, 28, 30; Ward's Communications 2003, 2, 47, 49; Sheehan 2001, 13.

16. This estimate of the costs of traffic collisions is by the National Highway Safety Administration, as cited in Thomas 2004, 26. In recent years, around 43,000 have been killed and 2.6 million injured annually in traffic collisions in the United States.

17. Foster 1981, 14.

18. Yago 1984, 59–61; Freund and Martin 1993, 135–137; Dunn 1998, 8; Paterson 2000, 268.

19. Freund and Martin 1993, 136.

20. Based on data in Ward's Communications 2003, 14, 47–49. A large literature exists on the growth of the auto industry. See, for example, Sinclair 1983; Yang 1995.

Chapter 4

1. The Clean Air Act was amended and strengthened once again in 1990.

2. Madsen 1995, 207–234; McAuley 2003, 5414. These advances in the environmental performance of vehicles were part of a broader trend in the developed world after the 1960s to strengthen environmental and consumer protection (Vogel 2003, 557).

3. See California Air Resources Board, "California's Air Quality History Key Events," at http://www.arb.ca.gov/.

Political scientist David Vogel (1995, 6) labels "the critical role of powerful and wealthy 'green' political jurisdictions in promoting a regulatory 'race to the top' among their trading partners" the "California effect" because this state "has been on the cutting edge of environmental regulation, both nationally and globally, for nearly three decades." See chapter 8 of his book for a comprehensive analysis of why and under what conditions the California effect occurs.

4. Ananthaswamy 2001, 18.

5. Manufacturers of Emission Controls Association (MECA), "Clean Air Facts—Motor Vehicle Emission Control: Past, Present, and Future," at http://www.meca.org/. Given recent regulations and technological trends, MECA estimates that, by 2009, "automobiles sold in the U.S. will be 99 percent less polluting than vehicles sold in the 1960s." See MECA, "Clean Air Facts—Motor Vehicle Emissions and Air Quality in the U.S.," at http://www.meca.org/.

6. California Air Resources Board, "California's Air Quality History Key Events," at http://www.arb.ca.gov/.

7. Environment Canada, "Gasoline," and "Vehicles, Engines and Fuels: Scrappage Program," at http://www.ec.gc.ca/. For details on Air Care in Vancouver, see http://www.aircare.ca/.

8. The end-of-life vehicles directive is careful with its terminology: "reuse" refers to when a car part is used another time for the same purpose; "recycling" to when a car part is reprocessed into another original or alternative part; and "recovery" to when the energy in a car part is recovered through, say, incineration (Brinkler 2004, 9). The use of these terms by automakers, however, is inconsistent. Although some are doing so inadvertently, others are intentionally reporting incineration as "recycling" or "reuse" (Ecology Center 2005, 6).

9. Fenton 2000, 40.1; the Steel Recycling Institute, "Steel Recycling Fact Sheet," and "Recycling Scrapped Automobiles," at http://www.recycle-steel.org/.

10. Bellmann and Khare 1999, 721, 724; Bandivadekar et al. 2004, 22.

By weight, glass constitutes about 3 percent of a typical vehicle. Some progress has been made, especially in Japan and Europe, in recycling automobile glass since the late 1990s. Many hurdles still exist, however, to effective and efficient glass reprocessing. These include contamination of the glass waste stream with metals and plastics as well as with mixes of different types of glass, which can affect coloring and quality (Brinkler 2004, 10–11).

11. Ecology Center 2005, 2.

12. Schaffer 2004, 4; Bandivadekar et al. 2004, 2223; Bellmann and Khare 1999, 721–733.

13. Ecology Center 2005, 2–3. Despite its praise, the center gave Toyota only a "C."

14. WHO and World Bank 2004, 132–133.

15. "Safety: Improving Your Odds," *Consumer Reports* 68 (April 2003), 26; WHO and World Bank 2004, 91–92.

16. National Highway Traffic Safety Administration data, as summarized in Cloud 2003; Department of Transportation data, as summarized in Fonda 2004, 65.

17. WHO and World Bank 2004, 37; "Safety: Improving Your Odds," *Consumer Reports*, April 2003, 26.

Chapter 5

1. The estimated current number of passenger cars and commercial vehicles extrapolates from 2001–2002 trends in Ward's Communications 2003, 49 (plus the latest figures released by Ward's Communications). The data on vehicles registered globally are from AMA 1951, 28; AAMA 1994, 41; Ward's Communications 2002, 53.

2. "Perpetual Motion," *Economist*, 4 September 2004, 4; "The Car Company in Front—Toyota," *Economist*, 29 January 2005, 73; Worldwatch Institute 2006; McAuley 2003, 5414.

3. Womack, Jones, and Roos 1990.

4. "The Car Company in Front—Toyota," *Economist*, 29 January 2005.

5. "London Congestion Charges No Barrier for Toyota Prius," *Management Services*, April 2003, 5; "Why the Future Is Hybrid," *Economist* 373, 4 December 2004, 22. Not all hybrids are saving significantly on energy. The 2005 Honda Accord hybrid got only 1 mile per gallon more than the four-cylinder Honda EX. The selling feature of the electric engine in the Honda Accord was, not better gas mileage, but better acceleration—from 0 to 60 miles per hour in just 6.9 seconds (compared to 9.0 seconds for the 4-cylinder Accord EX).

6. "The Car Company in Front—Toyota," *Economist*, 29 January 2005, 65–67.

7. Worldwatch Institute 2004a, 6.

8. McAuley 2003, 5415; Freund and Martin 1993, 19.

9. Expensive and strict inspections for motor vehicles older than three years encourage Japanese consumers to trade in cars more quickly than in Europe or North America. So do the high costs of maintenance, repairs, and parts. These economic factors, along with a general consumer preference for new products, explain the relatively low demand within Japan for used cars. On the other hand,

stronger local demand for them in Europe and North America creates fewer incentives to export used cars from these regions.

10. McAuley 2003, 5414; Worldwatch Institute 2004a, 6; Ward's Communications 2003, 14; Yardley 2004, A4; and General Motors at http://www.gm .com/. See Gallagher 2006 for a book-length analysis of China's increasing role in shaping how automobiles affect global environmental change.

11. "Perpetual Motion," *Economist*, 4 September 2004, 4; UNEP 2002, 35; Freund and Martin 1993, 27.

12. UNCHS 2001, 68; Gordon, Mackay, and Rehfuess 2004, 28. The Ontario Medical Association study is summarized in Rutledge 2005, F7.

13. Myers and Kent 2004, 29; U.S. Department of Energy data, as summarized in McAuley 2003, 5415.

14. Ananthaswamy 2001, 18; Energy Information Administration 2007, 1, 5.

15. Union of Concerned Scientists, "Automaker Rankings: The Environmental Performance of Car Companies," at http://www.ucsusa.org/.

16. Andrew Goudie's conclusions are summarized in Vince 2004.

17. Bellmann and Khare 1999, 733.

18. See Chen 2005, 20–26.

19. Simms and O'Neill 2005, 787.

20. WHO and World Bank 2004, 3–4; Crandall, Bhalla, and Madeley 2002, 1145.

21. Based on United Nations statistics for road traffic accidents, as cited in Haegi 2002, 1110.

22. The National Safety Council's estimate of lifetime odds assumes a life expectancy of 78 years. See National Safety Council, "Resources," at http://www.nsc .org/.

23. WHO and World Bank 2004, 37.

24. Nantulya and Reich 2002, 1139–1140; WHO and World Bank 2004, 4–5.

Chapter 6

1. WHO and World Bank 2004, 3–4.

2. "The Car Company in Front—Toyota," *Economist*, 29 January 2005.

3. For an analysis of the resistance to automobiles, see McKay 1996; Wall 1999; Robinson 2000; Paterson 2007. For an analysis of the movement toward voluntary simplicity, see Maniates 2002b, 199–236. For a discussion of "culture jamming," see Bordwell 2002, 237–252. For a study of the Toronto Islands, see Princen 2005, chap. 8.

4. "What Would Jesus Drive?" at http://www.whatwouldjesusdrive.org/. For a book-length critique of SUVs, see Bradsher 2002.

Chapter 7

1. According to the 1927 Ethyl Gasoline Corporation pamphlet, "The Story of Ethyl Gasoline," the team tested 33,000 compounds. Although this estimate is sometimes repeated (for example, in Kauffman 1989), it's probably on the high side. It was in the financial interest of the corporation to market ethyl gasoline as a historic breakthrough following years of painstaking research. Midgley once put the number of tests at just below 15,000, although the team may have run far fewer (see Kovarik 1999, n49).

2. See the comments of George Otis Smith (1920), director of the U.S. Geological Survey.

3. Charles F. Kettering, as quoted in Kovarik 2003. The potential of industrial alcohol as an automotive fuel has a long history. Henry Ford, for example, ran his first car on farm alcohol.

4. Thomas Midgley Jr., as quoted in Kauffman 1989, 719.

5. Kauffman 1989, 719. The cost estimate of a "penny" is from Kovarik 2005, 385.

6. The Du Pont family, using profits from gunpowder sales during World War I to purchase shares in General Motors, owned about one-third of GM shares by 1920. Pierre du Pont (1870–1954) became GM president while his younger brother, Irénée, was head of DuPont. The two firms were cooperating during this time on developing a better fuel. See Bent 1925, 3; Kitman 2000, 17.

7. Thomas Midgley Jr. to H. S. Cumming, 30 December 1922, as quoted in Rosner and Markowitz 1985, 345.

8. Needleman 1998, 79; Markowitz and Rosner 2002, 17. The GM newspaper ad for "ethyl gas," published in 1923, is reproduced in Kauffman 1989, 721.

9. GM and Standard Oil (later called "Exxon") combined their patents to create the Ethyl Gasoline Corporation.

10. Bayway laboratory manager, as quoted in, "Odd Gas Kills One, Makes Four Insane," *New York Times*, 27 October 1924, 11.

11. "Bar Ethyl Gasoline as 5th Victim Dies," *New York Times*, 31 October 1924, 1.

12. Bureau of Mines' report, as excerpted in "No Peril to Public Seen in Ethyl Gas," *New York Times*, 1 November 1924, 17. That the report was distributed to the media just days after the Bayway deaths was not a coincidence. Under the terms of its contract with the Bureau of Mines, GM in effect controlled the release of the findings: the bureau had to submit drafts to Ethyl for "comment, criticism, and approval" (GM contract, as quoted in Rosner and Markowitz 1985, 345).

13. During the first year and a half, 15 to 20 workers likely died in the three plants producing tetraethyl lead. Some scholars arrive at a "precise" number for the estimate of deaths. Yet, for many reasons, it's prudent to settle for a range of "likely" deaths. Officials did not always accurately record the cause of death

for some workers. The complexities of diagnosis and corporate incentives to avoid blame could mean the number of deaths was higher—perhaps far higher. But it could have been lower instead. Friends and colleagues of Bayway worker Joseph G. Leslie, for example, believed he died in 1924. His wife knew the truth, however: doctors had confined him in a psychiatric hospital, where he would die 40 years later (Kovarik 2005, 384).

14. Bent 1925, 3.

15. Until May 1925, Ethyl had been making steady inroads into the U.S. gasoline market. In the first two years and two months, it had distributed some 300 million gallons of ethyl gasoline to 12,000 filling stations in 28 American states (Bent 1925, 3).

16. Frank Howard, as quoted in Rosner and Markowitz 1985, 348. For the full proceedings of the conference on the health effects of tetraethyl lead, see U.S. Public Health Service 1925.

17. Report of the Surgeon General's committee of experts, as quoted in Kauffman 1989, 721. The Surgeon General qualified the evaluation, however, with a call for further independent research. For the full report, see U.S. Surgeon General 1926.

18. Kauffman 1989, 721; see also Kovarik 2003.

19. Alfred P. Sloan, as quoted in Kovarik 1999, n116; see also Kitman 2000, 31.

20. See the National Inventors Hall of Fame entry for Thomas Midgley Jr., at http://www.invent.org/.

21. Kauffman 1989, 722.

22. Professor Yandell Henderson, speech to the American Society of Safety Engineers and International Safety Council, as excerpted in "Sees Deadly Gas a Peril in Streets," *New York Times*, 22 April 1925, 25.

23. David Edsall, as quoted in Kitman 2000, 31. Alice Hamilton, as quoted in Kovarik 2005, 388–389. Erik Krause to Thomas Midgley, as quoted in Kitman 2000, 20, and Kovarik 2005, 385.

24. H. C. Parmelee, as quoted in "Demands Fair Play for Ethyl Gasoline," *New York Times*, 7 May 1925, 10. The figure on the market share of leaded gasoline by the 1930s is posted on the Web site of Professor William Kovarik of Radford University, http://www.radford.edu/.

25. Graebner 1987: 140–159; Needleman 1998, 80. See also Markowitz and Rosner 2002.

26. The new standard of 4.23 grams of tetraethyl lead per gallon of gasoline was, again, the outside limit of what refiners were using. The average for most refiners during the 1950s and the 1960s was about 2.4 grams per gallon (see Lewis 1985).

27. Surgeon General, as quoted in Lewis 1985.

28. Patterson 1965, 344, 358. Patterson's later research on the Arctic, Antarctic, oceans, freshwater, terrestrial soils, and food chains would add copious evidence

of a global buildup of lead from human sources. He received many honors during his life (1922–95), including the naming of an asteroid and Antarctic mountain peak after him. See Flegal 1998 for a summary of Patterson's influence on environmental research. See Davidson 1999 for a collection of essays in tribute to his research.

29. Clair Patterson and Robert Kehoe, as quoted in Needleman 2000, 22–23, which also has excerpts from Kehoe's correspondence with Katharine R. Boucot, editor of the *Archives of Environmental Health*.

30. Portions of Kehoe's testimony are reprinted in Needleman 2000, 24–25. Needleman (1998, 79–85) and Nriagu (1998, 71–78) compare the views of Patterson and Kehoe.

31. Needleman 2000, 26.

32. Gould 1962, 1.

33. Joseph C. Robert (1983), as quoted in Needleman 2000, 26. Robert's assessment of GM may well be unfair. We know now that GM first began developing the catalytic converter in 1958—four years before selling Ethyl (Kovarik 2005, 394).

34. Rawleigh Warner Jr., as quoted in Abele 1970, 51.

Chapter 8

1. Thomson 2000, 186. Many industry scientists in the 1970s contested estimates such as "around 80 percent" for the contribution of automobiles to airborne lead. My confidence in this figure arises from later studies, including a comprehensive review, which found automotive exhaust was the "greatest source of atmospheric lead pollution" (Lansdown and Yule 1986, 137).

2. William Ruckelshaus, as quoted in Markowitz and Rosner 2002, 117.

3. Chair of the National Academy of Sciences review team, as quoted in Needleman 2000, 28.

4. U.S. EPA 1973.

5. Panel of the U.S. Court of Appeals for the District of Columbia, as quoted in Kitman 2000, 37; see also Kovarik 2005, 394.

6. The 1973–74 newspaper ad is described in Needleman 2000, 28, and in Thomson 2000, 188.

7. U.S. Court of Appeals for the District of Columbia, as quoted in Thomson 2000, 194.

8. U.S. EPA data, as summarized in Needleman 2000, 30–32.

9. Thomson 2000, 190, 200.

10. The International Lead Zinc Research Organization was also continuing to support research. The *Proceedings* of the Second International Symposium on Environmental Lead Research, held in Cincinnati in 1978, provides some revealing examples; see Lynam, Piantanida, and Cole 1981.

11. Needleman 2000, 31–32; Thomson 2000, 188; Lewis 1985.

12. Lawrence Blanchard, as quoted in Needleman 2000, 31.

13. Donald R. Lynam, Ethyl's director of air conservation, as quoted in Noble 1984, E8; Ashland Oil's position is summarized in Thomson 2000, 189.

14. Indeed, the EPA science advisory council went on to praise Needleman's methods and results as "pioneering." For details on the controversy surrounding Herbert Needleman, see Markowitz and Rosner 2002. For an example of one of his most controversial articles, see Needleman et al. 1979.

15. Newell and Rogers 2004, 178–181.

16. The estimate of 99.8 percent for the decrease in lead content of gasoline from 1976 to 1990 is in Ostro 2000, 5. A summary of the U.S. Department of Health and Human Services survey is in Kitman 2000, 38.

Until the 1960s, most doctors put the "safe" threshold for lead at 60 micrograms per deciliter of blood. The reason was straightforward: visible symptoms of acute lead poisoning, such as convulsions, generally do not occur below this level. Brain and kidney damage can occur at lead levels between 80 and 100 micrograms per deciliter of blood. Death can result at levels above 125 micrograms per deciliter of blood (Ostro 2000, 2).

17. Needleman 2000, 34. Needleman's estimate assumes toxic levels are above 15 micrograms per deciliter of blood, the government standard in 1988. For the estimate of declining lead levels in children from 1976 to 1994, see Centers for Disease Control and Prevention 2005, 513.

18. The Japanese government was one of the first to act, phasing down leaded gasoline in the 1970s. By 1981, only 3 percent of gasoline in Japan contained lead; five years later, no one was producing or using leaded gasoline there (El-Fadel and Hashisho 2001, 38).

19. Ethyl's data are summarized in Kitman 2000, 13, 39.

20. Hilton 2001, 248.

21. El-Fadel and Hashisho 2001, 38.

22. For an analysis of Thailand's phaseout of leaded gasoline from 1991 to 1995, see, for example, Sayeg 1998.

23. "Getting the Lead Out," *Time Canada*, 27 July 1998, 12. (The article summarizes data from the World Bank and World Resources Institute.) The country percentages hide some significant variations in environmental impacts arising from different numbers of vehicles and amounts of lead in the gasoline pool. At the end of the 1990s, for example, automobiles were emitting on average about 20 times more lead per day in Mexico City than in Jakarta (Brooke 2000, 27).

24. Hammar and Löfgren 2004, 192–205. See also Löfgren and Hammar 2000, 419–431.

As in the United States, many refiners and gasoline station owners in Europe opposed the phaseout of leaded gasoline because of the costs of introducing unleaded gasoline. Smaller operators in particular argued that any phaseout would favor larger operators—including ones with ties to the United States.

25. See Hammar and Löfgren 2004, 192–193, 203.

26. Hilton 2001, 246–247. Hilton analyzed 19 countries with phaseouts completed by 1994. The mean phaseout time was about 15 years for those countries starting before 1979. The mean time for those beginning later was about 10 years. The one significant exception was Japan, which, although one of the first to begin phasing out lead, largely did so in just 4 years. This analysis excludes countries that did not manage to phase out lead from gasoline by 1994, leaving open the possibility that some countries began after 1979, but still took longer than countries like the United States. See Hilton 2001, 249–53, for his cases and analysis.

27. By the late 1990s, Octel was supplying 80 percent of the world's tetraethyl lead, including to companies like the Ethyl Corporation (Brooke 2000, 27). The NewMarket Corporation is now the parent company of the Ethyl Corporation. Its Web site (http://www.ethyl.com) explains the company's ongoing link to leaded gasoline: "Ethyl also has a significant interest in the tetraethyl lead (TEL) business through marketing agreements with Innospec Inc. This sunset business continues to generate strong cash flow for the company."

28. Member of Octel's management board, as quoted in Kelly and Dawley 1995, 50.

Chapter 9

1. See "Declaration of Dakar, Regional Conference on the Phasing-Out of Leaded Gasoline in Sub-Saharan Africa," at http://www.unep.org/; Lacey 2004b, 5.

2. "Lead-Free Africa, Lead-Free World," *Appropriate Technology* 30, no. 1 (March): 26; a summary of Robert De Jong's comments is in Majtenyi 2005.

3. Phiri 2006; Lacey 2004a, N3.

4. Timberg 2006, A14; Lacey 2004a, N3.

5. UNEP 2004; "House Team Roots for Oil Refinery Upgrade," *East African Standard*, 6 February 2006; "Kenya Phases Out Leaded Fuel," *East African Standard*, 27 January 2006; "Leaded Petrol May Be Phased Out This Month," *Nation*, 7 January 2006.

6. Worldwide, different brands of lead replacement petrol (LRP) rely on different additives. Thus the LRP in countries like South Africa generally contains potassium, phosphorus, or manganese.

7. UNEP 2005b; Timberg 2006, A14; Phiri 2006. The estimate of "at least 600,000 children" is by the South African Medical Research Council.

8. Majtenyi 2005; Timberg 2006, A14; Phiri 2006.

9. Over 30 countries worldwide were still using leaded gasoline at the beginning of 2006; among these, countries without any phaseout plans included Afghanistan, Algeria, Bhutan, Burma, Cambodia, Cuba, Iraq, Laos, Mongolia, North Korea, Tajikistan, Turkmenistan, and Uzbekistan.

Chapter 10

1. Wakefield 2002, A574–A580; Ostro 2000, 1–30.

2. In the case of the United States, the overall benefits of the phasedown of leaded gasoline "likely outweighed its costs by 10 to 1" (Newell and Rogers 2004, 191).

3. Newell and Rogers 2004, 189.

4. Manufacturers of Emission Controls Association (MECA), "Clean Air Facts—Motor Vehicle Emission Control: Past, Present, and Future; and The Motor Vehicle Emission Control Industry," at http://www.meca.org/.

5. Midgley's associate Carroll Hochwalt, as quoted in Cagin and Dray 1993, 11. For an obituary of Thomas Midgley Jr., see also *Time*, 13 November 1944, 86; for a flattering biography, written by Midgley's grandson, Thomas Midgley IV, see Midgley 2001.

Chapter 11

1. Paul Crutzen of the Max Planck Institute for Chemistry in Mainz, Germany, was awarded the 1995 Nobel Prize in Chemistry along with Molina and Rowland. As early as 1970, Crutzen had raised the possibility that nitrogen oxides from fertilizers and supersonic aircraft were harming the ozone layer.

2. Cagin and Dray 1993, 58; see also Kelly 2005, 52.

3. Cagin and Dray 1993, 58–59.

4. The widespread use of similar compounds in the CFC class came later. CFCs were initially put into insecticide spray cans to propel insecticide for combatting malaria during World War II. After the war, this application was extended to a variety of other products, from hairsprays to deodorants. In the 1950s, CFCs were mixed with plastic resin to create polyurethane, common in insulation, car bumpers, egg cartons, and picnic coolers. The use of CFCs in air conditioners grew quickly after the 1950s with sales of air-conditioned automobiles and the building of air-conditioned malls, office towers, and sports arenas. Later still, CFCs became common in solutions to clean computer chips and electronic parts.

5. DuPont bought out GM's share in 1949. Before long, other CFC producers were entering the market, including Allied Signal in 1952, Pennwalt in 1957, Kaiser Tech in 1963, and Racon in 1965. See Thévenot 1979 for a comprehensive history of refrigeration.

6. Cagin and Dray 1993, 59, 66–67; see also Kelly 2005, 52.

7. Molina and Rowland 1974. Their article sparked a burst of research; their original theory was refined as more and more evidence was gathered.

8. Two of the most common CFCs were chlorofluorocarbon-11 and -12.

9. F. Sherwood Rowland, as quoted in "Chlorofluorocarbons Threaten Ozone Layer," 168th American Chemical Society National Meeting, *Chemical and Engineering News*, 23 September 1974, 28; see also Rowland 2000, 137.

10. Molina and Rowland's press conference is summarized in Cagin and Dray 1993, 6–9. See also "Chlorofluorocarbons Threaten Ozone Layer," 168th American Chemical Society National Meeting, *Chemical and Engineering News*, 23 September 1974, 27–30; "Environmentalists Seek Fluorocarbon Ban," *Chemical and Engineering News*, 2 December 1974, 14; "Fluorocarbon Ban Would Be Premature," *Chemical and Engineering News*, 23 December 1974, 12.

11. Shapiro 1975, 13.

12. NOAA, as quoted in Gibney 1975, 13.

13. Rowland 2000, 137.

14. See the DuPont Company statement, "You Want the Ozone Question Answered One Way or the Other; So Does Du Pont," *Washington Post*, 30 September 1975, A10.

15. Strictly speaking, the "hole" in the ozone layer was an area of severe thinning. In 1985, some 45 percent of CFCs produced in the United States were used as refrigerants, 30 percent as blowing agents, and 20 percent as cleaning agents. For background on developments in the early to mid-1980s, see Ciantar and Hadfield 2004.

16. In the mid-1980s, halons (whose molecules combine carbon with one or more atoms of bromine) were common in items like fire extinguishers. Although less significant than CFCs, halons deplete the ozone layer, too.

17. DuPont began investigating possible substitutes in the mid-1970s, but, according to the manager of DuPont's Freon division in 1988, it ended active testing in 1980 after only managing to develop substitutes that were too costly for most consumers. Summarized in Weisskopf 1988, A1.

18. Calculated from data on the Alternative Fluorocarbons Environmental Acceptability Study (AFEAS) Web site, http://www.afeas.org/. These data refer only to the five main CFCs: CFC-11, CFC-12 (the primary one for refrigerators), CFC-113, CFC-114, and CFC-115. See Litfin 1994 and Parson 2003 for analysis of the global environmental politics of developing an international agreement to address ozone depletion.

19. Parson 2003, vii.

Chapter 12

1. Production of CFC-113 climbed from just over 100,000 metric tons in 1980 to a peak of just over 250,000 metric tons in 1989. The phaseout was swift after the Montreal Protocol went into force, with global production falling off to just 6,000 metric tons in 1996. See the data on the Alternative Fluorocarbons Environmental Acceptability Study Web site, http://www.afeas.org/.

2. UNEP 1989, iii, 7–8. This report estimated the total amount of controlled CFCs in 1986 at approximately 1.1 million metric tons.

3. Ahmed 1995, 31.

4. Electrolux 2005, 32; BSH 2004, 21. See Langley 1994 for background on the use and recycling of CFCs in the refrigeration sector.

5. Hammitt 2004, 172. For details on the illegal trade in CFCs, see Clapp 1997. For more on the history of regulating CFCs, see Benedick 1998; Andersen and Sarma 2002.

6. Zambian fishmonger and civil servant, as quoted in Zulu 1999, 6.

7. "CFC Refrigerators to be Phased Out in India," *New Nation*, 14 January 2005. Many countries in Eastern Europe also converted the cooling systems of CFC refrigerators to run on hydrocarbons. China is also using this technology.

8. Zhao 2005, 59; Zhao and Ortolano 2003, 710; Zhao and Ortolano 1999, 500–503.

9. See Zhao and Ortolano 2003, 708–725; Zhao 2005, 58–81.

10. Chinese refrigerator plant manager, as quoted in Zhao and Ortolano 1999, 503.

11. Zhao and Ortolano 1999, 503–505. For details on environmental labeling in China's refrigeration sector, see Zhao and Xia 1999, 477–497.

12. In 2004, BSH, for example, became the first company to produce CFC- and HFC-free refrigerators in Brazil at its subsidiary's Hortolândia factory near São Paolo (BSH 2004, 21).

13. Zhao and Ortolano 1999, 504, table 1.

14. Complex, and at times frustrating, application procedures for Multilateral Fund financing slowed plans to phase out CFCs and halons across all sectors in China using them in the 1990s. It was especially hard for smaller firms to meet the fund's criteria. Since 1997, financing packages from the fund for entire sectors, rather than individual firms, have helped overcome some of these procedural obstacles (Zhao 2005, 67–68).

15. Zhao 2005, 63–64, 70.

16. See Multilateral Fund, "News, Multilateral Fund Looks to 2007 and 2010 Targets," 11 April 2006, at http://www.multilateralfund.org/.

17. Skaer 2001, 106.

18. See Greenpeace, "Greenfreeze and Solar Chill" at the Greenpeace Web site, http://www.greenpeace.org/.

19. Global production of hydrochlorofluorocarbons (HCFC-22, -124, -141b, and -142b) in 2003 was approximately 284,000 metric tons, down from a peak of about 445,000 metric tons in 1998. The trend for hydrofluorocarbons (HFC-134a, -125, and -143a) has been in the other direction, however, with global production increasing from about 113,000 metric tons in 1998 to over 202,000 metric tons in 2003. See the Alternative Fluorocarbons Environmental Acceptability Study Web site, http://www.afeas.org/, for production data on HCFCs and HFCs from 1970 to 2003.

20. In 1993, for example, the Sacramento Municipal Utility District offered consumers $100 for each old (CFC) refrigerator they replaced with a new (CFC-free) one (Kim, Keoleian, and Horie 2006, 2311).

21. The data on refrigerators and freezers are from Euromonitor International 2003, 5.1.4 Regional Sales.

22. Euromonitor International 2003, 8.1 Corporate Trends.

23. China Sourcing Reports 2005.

24. Euromonitor International 2003, 8.1 Corporate Trends.

25. Consumers take on average between 11 and 13 years to replace refrigeration appliances (an approximate global figure, with moderate variations across particular countries). Euromonitor International 2003, 3.4.2 Refrigeration Appliances.

26. The increasing interest in hobby cooking in the United States has also pushed up demand for new (CFC-free) appliances (in part because some celebrity chefs encourage such upgrades).

27. Euromonitor International 2003, 5.1.5 Leading National Markets.

Chapter 13

1. Kim, Keoleian, and Horie 2006, 2310; Euromonitor International 2003, 3.7.1 Energy Efficiencies; Electrolux 2005, 9.

2. Kim, Keoleian, and Horie 2006, 2310; Higgins 2001, 50.

3. BSH 2004, 23; BSH 2006a, 4.

4. In 2002, the world's largest manufacturer of refrigeration appliances was the Swedish firm Electrolux (with 12.0 percent of the world share in unit volume); the U.S. firm Whirlpool was close behind, with 10.8 percent; the remaining manufacturers by unit volume were Bosch und Siemens Hausgeräte (BSH; 5.6 percent); Haier Group (5.6 percent); General Electric (4.6 percent); and Maytag (3.0 percent). See Euromonitor International 2003, 5.1.8 Global Manufacturer and Brand Shares. BSH grew over the next few years and, by 2006, was capturing 7 percent of the global home appliance market (BSH 2006a, 13).

5. Electrolux 2005, 5.

6. The International Organization for Standardization has also certified seven nonmanufacturing Electrolux units under ISO 14001.

7. Electrolux 2005, 8–9, 24, 33. Henrik Sundström is quoted on p. 8.

8. Electrolux 2005, 5, 9, 17, 25–27. Jean-Michel Paulange is quoted on p. 27. See also the Web site for the World Packaging Organization, http://www .packaging-technology.com/wpo/.

9. The WEEE Directive is a "minimum" directive, meaning significant differences can arise among the national laws of member states.

10. Electrolux 2005, 10–11. In 2001, Japan also made manufacturers legally responsible for dismantling and recycling large home appliances. Japanese consumers must pay a set fee to throw out an appliance. No plans are under way in North America for similar recycling laws, primarily because, thanks to the high steel content of appliances, the recycling rate is already high (85 percent

in 2001 in the United States). See Euromonitor International 2003, 3.7.4 Recycling/Disposal.

11. The six hazardous substances restricted by the European Union are cadmium, hexavalent chromium, lead, mercury, polybrominated biphenyl (PBB), and polybrominated diphenyl ether (PBDE). See Selin and VanDeveer 2006 for an overview of the EU's recent management of hazardous substances.

12. BSH 2004, 5. See also BSH 2006b and the BSH Web site, http://www .bsh-group.com/. *Environmental and Corporate Responsibility 2006* marked the fifteenth straight year that BSH published an environmental responsibility report.

13. The CECED brings together about 280 of Europe's major home appliance makers, which together produce about 50 million large appliance units in Europe every year. The CECED code of conduct (see CECED's Web site, http://www .ceced.org) reads: "Companies will comply with environmental regulations and standards applicable to their operations, and will observe environmentally conscious practices in all locations where they operate" (p. 3).

14. The high estimate (90 percent) is the one BSH uses to calculate its environmental impacts (BSH 2006a, 4). The low estimate (80 percent) is in Electrolux's sustainability report (Electrolux 2005, 8).

Transportation of an appliance, for example, has relatively little environmental impact compared to its use. Trucking a refrigerator 3,000 kilometers (1,900 miles) from Sweden to Spain generates 14 kilograms of carbon dioxide. This is equal to about 30 days of use of a top energy class refrigerator or freezer (Electrolux 2005, 33).

15. BSH 2004, 9, 20, 23–26, 30.

16. See "Corporate Fact Sheet," at http://www.whirlpoolcorp.com/. Globally, Whirlpool expects unit demand to continue to grow at a rate of between 2 and 3 percent, and for the company to therefore continue to expand operations (Whirlpool Corporation 2006, 4).

17. See "Social Responsibility: Our Commitment to Corporate Responsibility," on the corporate Whirlpool Web site, http://www.whirlpoolcorp.com/.

18. ENERGY STAR 2006, 1–2.

19. See "About ENERGY STAR," at http://www.energystar.gov/. See also ENERGY STAR 2006.

20. ENERGY STAR 2006, 1–2; "EPA Recognizes Energy Star Winners for Outstanding Energy Efficiency," 21 March 2007, in the "News Room," at http://www.energystar.gov/.

21. See Built Green Colorado 2006.

22. See "EPA Recognizes Energy Star Winners for Outstanding Energy Efficiency," 21 March 2007, in the "News Room," at http://www.energystar .gov/.

23. Myers and Kent 2004, 54–55.

24. Kim, Keoleian, and Horie 2006, 2310 (based on data from the U.S. Association of Home Appliance Manufacturers).

25. Energy Information Administration 2005, 1; Myers and Kent 2004, 54. The 2.2 percent estimate for rising annual energy consumption since 1970 is for "marketed" energy.

26. Energy Information Administration 2005, 1.

27. Economist Intelligence Unit 2005, 7, 21. For more on per capita energy consumption, see International Energy Agency 2006. The United States has slightly higher per capita carbon dioxide emissions than countries like Canada and Australia, far higher per capita emissions than countries like Germany, the United Kingdom, and France, and dramatically higher than most of the developing world. Here is a sample of the figures in metric tons per capita for 2004: United States (20.2), Australia (19.4), Canada (18.1), Germany (10.5), Japan (9.9), United Kingdom (9.6), France (6.7), Mexico (3.7), China (3.6), India (1.0), and Rwanda (0.1). Singapore, interestingly, has some of the world's highest per capita emissions (29.7). These statistics are from the U.S. Energy Information Administration, at http://www.eia.doe.gov/.

28. The amount of carbon dioxide in the atmosphere has increased by 35 percent since the late 1700s, with the increase attributed primarily to the combustion of fossil fuels (World Meteorological Organization 2006, 2).

29. The 1990–2001 emissions data, from the U.S. Energy Information Administration, are in Myers and Kent 2004, 53. The Netherlands Environmental Assessment Agency conducted the study placing China as the world's largest emitter of carbon dioxide in 2006 (see http://www.mnp.nl/en).

Chapter 14

1. The ozone hole over the Antarctica broke records in September 2006 for depth, average area, and duration (9 days, from September 21 to 30). Still, the overall signs are positive. The ozone layer has remained fairly stable since the mid- to late 1990s, despite some scientists' concern that its decline might be impossible to halt. The reasons for changes in the ozone layer are complex, including sunspots, weather, and volcanoes; some fluctuations are natural. Reduction in the emission of CFCs probably accounts for about half of the layer's recent stability. See Barry and Phillips 2006. See also UNEP's Web site, http://www.unep.org/, and NASA's Web site, http://www.nasa.gov/, for the latest data.

Some scientists do worry, however, that climate change may derail a full recovery of the ozone layer (see Weatherhead and Andersen 2006, 39–45).

2. In total, developing countries were able to reduce CFC consumption by 60 percent from the mid-1990s to 2004 (UNEP 2005a).

3. Higgins 2001, 54.

Chapter 15

1. Sinclair's *The Jungle* first appeared in 1905 as a serial in the socialist weekly *Appeal to Reason*. An edited version (with a modified ending) was first published as a novel in 1906 by Doubleday, Page. The quote from Sinclair's autobiography is in Sinclair (1962), reprinted in Eby 2003, 351.

2. The efforts in Congress to regulate food production began years before the publication of Sinclair's novel. The "scandal" from this novel, however, "shocked the world" and galvanized public support (Eby 2003, ix).

3. Although Henry Ford is correctly remembered as the first to use the moving assembly line for large-scale manufacturing (building entire factories around the idea), James Barrett (1987) documents the meatpackers of Chicago as the first to use it at all.

4. Smil 2002, 606. For a sampling of the literature on the history of farming and eating beef, see Rixson 2000; Carlson 2001; Smil 2000, 2002; Rogers 2003.

5. Leading authority Vaclav Smil of the University of Manitoba summarizes these studies in Smil 2002, 606–610. No firm consensus exists on the best estimates of average per capita meat consumption during the Middle Ages. Although some scholars of European history accept much higher figures than the ones in this chapter (such as 500 kilograms in Berlin in 1397 and 72–100 kilograms in Nuremberg in 1520), Professor Smil rejects these as "vastly exaggerated."

6. Skaggs 1986, 90; Smil 2002, 609. For details, see also Popkin 1993; Caballero and Popkin 2002.

7. Gold 2004, 8; Brown 2006, chap. 9; Nierenberg 2005, 9–10. Sheep, goats, and other less common animals like buffaloes, yaks, and ducks account for the remaining 7 percent of global meat production.

8. The data for China, India, Indonesia, and the United States are from the FAO, World Resources Institute, "Earthtrends," at http://earthtrends.wri.org/.

9. U.S. Department of Agriculture (USDA) statistics, from the "Food Safety and Inspection Service Fact Sheets," at http://www.fsis.usda.gov/.

The federal government began in 1985 to charge a mandatory $1 per head of cattle at the time of sale, in part to fund generic advertising to promote beef consumption. The goal of the so-called Beef Checkoff Program is "to strengthen the position of beef in the marketplace and to maintain and expand domestic and foreign markets and uses for beef and beef products." For years, this program was under attack on various legal fronts as unconstitutional for amounting to "compelled speech." The U.S. Supreme Court finally ruled the program constitutional in 2005. Today, these fees generate about $45 million a year. For details, see "Beef Checkoff Program, Beef Promotion and Research Order," at http://www.ams.usda.gov/.

10. See Economic Research Service (ERS) of the U.S. Department of Agriculture, at http://www.ers.usda.gov/.

11. See FAO, "World Agriculture 2030: Main Findings," at http://www.fao.org/.

12. Nierenberg 2005, 12.

13. Smil 2002, 618; see also "Breeds of Cattle," at http://www.cattle-today.com/.

14. Schlosser 2001, 136–138; Goldenberg 2004, 13.

15. Schlosser 2001, 138. The estimate of the degree of control of the four largest U.S. meatpackers is from Hendrickson and Heffernan 2005.

16. Schlosser 2002, 26.

17. The United States, for example, allows ranchers to use antibiotics to "treat" or "prevent" diseases in cattle but requires a minimum waiting period before slaughtering to allow these to leave the animal's system. The Food Safety and Inspection Service of the USDA conducts random tests for antibiotic residues at slaughter time. The U.S. government also allows ranchers to use hormone implants (placed in an ear of each animal) to "promote efficient growth."

18. Nierenberg 2005, 23; Smil 2002, 616. The estimate of the percentage of cattle receiving growth hormones is on the Niman Ranch Web site, http://www.nimanranch.com/, under "Frequently Asked Questions."

The "grain" for animal feed—especially corn—is often of lower quality than the "grain" grown for human consumption. This is arguably an "efficient" use of "grain" (a case the meat industry sometimes makes in response to critics who see meat as an inefficient source of protein for humans).

19. These weight-gain estimates were made by rancher Rich Blair during an interview with journalist Michael Pollan (2002), 47.

20. See Schlosser's bestseller *Fast Food Nation* (2001) for a lively account of these modern slaughtering facilities in the United States. For a sampling of more academic studies on the political economy of beef production (from a variety of disciplines and countries), see Gouveia and Juska 2002; Brown, Longworth, and Waldron 2002; Filho 2004.

21. Schlosser 2002; Moss, Oppel, and Romero 2004, A1.

22. See Moss, Oppel, and Romero 2004, A1.

23. South Dakota rancher, Ed Blair, as quoted in Pollan 2002, 47.

24. See Schlosser 2001; McDonald's 2005 financial statements, at http://www.mcdonalds.com/; Subway, "About Us," at http://www.subway.com/; "About Pizza Hut," at http://www.pizzahut.com/; and Kentucky Fried Chicken, "About Us," at http://www.kfc.com/.

25. Wendy's spokesman, Denny Lynch, as quoted in Warner 2006, C5.

26. Lemonick and Bjerklie 2004, 58–69; World Health Organization, "Facts, Obesity and Overweight," at http://www.who.int/.

27. FAO data, World Resources Institute, "Earthtrends," http://earthtrends.wri.org/; Nierenberg 2005, 10–11. See also the Web site of the International Food

Policy Research Institute (a research center of the Consultative Group on International Agricultural Research), at http://www.ifpri.org/.

Chapter 16

1. Smil 2002, 609; FAO, "World Agriculture 2030: Main Findings," at http://www.fao.org/. See also the data tables of the World Resources Institute, under "Earthtrends," at http://earthtrends.wri.org/. For a general ecological argument against the use of grain to produce meat, see Goodland 1998.

Significant differences exist in how much cattle ranchers rely on grain. Those in developed countries, for example, tend to use far more grain to produce a kilogram of beef than those in developing ones (on average, around 9 times as much). This estimate comes from the Council for Agricultural Science and Technology, cited by the beef industry on its Web site, http://www.beef.org/.

2. Nierenberg 2005, 24; Millstone and Lang 2003, 35.

3. For background on increasing agricultural productivity after 1950, see Brown 2006, chap. 9.

4. This estimate of total 1998–2004 subsidies for U.S. soybean farmers is from the U.S. Department of Agriculture, summarized in Lawrence 2006, 8.

5. Ash, Livezey, and Dohlman 2006, 3, 11; Brown 2006, chap. 9.

6. Lawrence 2006, 8.

7. Economic Research Service of the USDA, "Soybeans and Oil Crops," in "Briefing Rooms," at http://www.ers.usda.gov/. Palm oil, currently the world's second largest source of vegetable oil, is beginning to close the gap on soybean oil with the rapid expansion in recent years of oil palm plantations in developing countries like Indonesia.

8. Economic Research Service of the USDA, "Soybeans and Oil Crops: Background," in "Briefing Rooms," at http://www.ers.usda.gov/; Oliveira and Davis 2006, 6.

9. Lawrence 2006, 8.

10. Kansas staff veterinarian, Mel Metzen, as quoted in Pollan 2002, 51.

11. David Wallinga, as quoted in Nierenberg 2005, 48 (see also pp. 32, 46–49); World Health Professions Alliance 2001.

12. Methane traps heat about 20 times—and nitrous oxide about 300 times—more effectively than carbon dioxide.

13. *Livestock and the Environment* 1999, 2, 9; Subak 1999, 79–91; Nierenberg 2005, 24. The 18 percent estimate for greenhouse gases is from a 2006 FAO report cited in Read 2007, C7. See also Worldwatch Institute 2004a.

14. The term "Amazon" in this section refers to the Brazilian government's administrative region, the "Legal Amazon."

15. See Kaimowitz et al. 2004, 1–4, 9; the import figures for fresh and frozen beef are calculated from the data in table 3 (p. 9).

16. David Kaimowitz, as quoted in "Making Mincemeat Out of the Rainforest," *Environment* 46, June 2004, 5; the information in this paragraph is drawn from Kaimowitz et al. 2004, 1–4.

17. See Kaimowitz et al. 2004, 2.

18. "Despite Foot and Mouth, Brazil's Beef Exports Break Record," *Brazzil Magazine*, 19 January 2006, at http://www.brazzilmag.com/.

19. These estimates of deforestation are from Brazilian National Institute of Space Research, reproduced on the Web site http://www.mongabay.com/.

Chapter 17

1. Montana cattleman, as quoted in Wilkinson 2003, 3.

2. See Coleman Natural Foods, at http://www.colemannatural.com/, in particular, "Frequently Asked Questions."

3. Dave Carter, former chairman of the USDA's National Organic Standards Board, as quoted in Moran 2006, C1. For details on the Food Safety and Inspection Service, see the U.S. Food Safety and Inspection Service, "Fact Sheets," at http://www.fsis.usda.gov/.

4. Natural beef producer, Charlie Moore, and Mel Coleman Jr., as quoted in Moran 2006, C1.

5. See Moran 2006, C1.

6. Burros 2006, F5. For background on grass farming, see http://www.eatwild .com/.

7. See the Niman Ranch Web site, http://www.nimanranch.com/.

8. Chef Heather Hand, as quoted in Robbins 2003, F8.

9. See the Niman Ranch Web site, http://www.nimanranch.com/. Organic farming in the United States accounted for just $393 million of the $207 billion in agricultural sales in 2002. That year, only 12,000 of the 2.1 million farmers were growing any certified organic products. See Purdum 2005, sec. 6, p. 76.

10. Robbins 2003, 8; see also the Niman Ranch Web site, http://www .nimanranch.com/.

11. The estimate of the retail value of organic beef in 2005 is from the U.S. Organic Trade Association. The estimate of the growth in sales of organic beef for that year is from the U.S. National Cattlemen's Association. See Aldrich 2006, B7C; Severson 2005, F1; Robbins 2003, 8; Roosevelt 2006, 76–78.

12. In the United Kingdom, for example, the Soil Association is the primary regulator of the organic sector, employing some 70 inspectors to approve licenses for farms. See Meikle 2006, 6.

13. German Chancellor Gerhard Schroeder announced that, "in the interest of the consumer," his government would work to increase the proportion of land farmed organically to 20 percent by 2010 (from 2.6 percent in 2001). See Sancton et al. 2001, 18–23.

14. Political scientist David Vogel (1995, 2003, 2004) documents many instances in more recent years where Europe has taken a more proactive tack than the United States in managing environmental risks.

15. Danish butcher and Parisian greengrocer, as quoted in Sancton et al. 2001, 18.

16. See Sancton et al. 2001, 18.

17. Ralph Human, managing director of the UK Organic Livestock Marketing Co-op, as summarized in Buss and Macmillan 2004, 30. See also Buss 2004, 29.

18. In 2005, the organic food market in the United Kingdom was growing at about twice the rate of the general grocery market. See Leigh 2005, 18.

19. Leigh 2005, 18.

20. Allison 2003, 39; Buss 2004, 29; Macmillan 2004, 30.

21. Burros 2002, F1.

22. Vancouver chef, as quoted in Gill 2002, L8.

23. See the Heifer International Web site, http://www.heifer.org/.

24. For details, see the Conservation International Web site, http://www.conservation.org/.

Chapter 18

1. Smil 2002, 606; Brown 2006, chap. 9; Nierenberg 2005, 9–10.

Chapter 19

1. Swilers would hunt a small number of hooded seals, too, ideally their pups, called "bluebacks," which produce more oil and larger pelts than whitecoats.

2. The "legend of the main patch," Newfoundland author Harold Horwood (1960, 38) believes, became part of sealing lore in the second half of the 1800s, as hunters searched for reasons why so many hunts failed to land large numbers of whitecoats. The obvious explanation—that hunters had severely depleted the herd with unsustainable catches in the 1830s and 1840s—was less attractive than searching for a "mythical main patch."

3. Candow 1989, 106. The term "highliner" originates from the tradition of placing the best fisherman in the bow, so that his fishing line was the highest in the water.

4. See Kean 1935, 13; see Cole 1997, 242–243, for a reprint of an article by N.C. Crewe, first published in the *Daily News* (St. John's), 31 March 1964; see Cole 1997, 245–249, for a reprint of an article by Mildred Gough, the daughter of Mary Crewe, first published in the *Gander Beacon*, 6 February 1985; see also Brown 1972, 1–2.

5. See Brown 1972 for a vivid account of this tragedy.

6. Chantraine 1980, 196. Captain Thomas Conners, who saw the *Southern Cross* steam "hell-bent out of the muck" of the storm across the stern of his ship *Portia*, felt the *Southern Cross* was overloaded with pelts, "half-seas under and so low in the water her decks was running green" (Mowat 1973, 136; the quote is a reenactment of Conners's recollections).

7. Chantraine 1980, 191; Candow 1989, 36; Brown 1972, 14. See Bartlett 1929 for a captain's description of sealing during the first quarter of the twentieth century.

8. The landsmen hunt—hunting from shore, inshore ice, and small boats—continued on a relatively small scale throughout the history of the large commercial sealing fleets.

9. Fisheries and Oceans Canada, "The Canadian Seal Hunt—A Timeline," at http://www.dfo-mpo.gc.ca/; Coish 1979, 16, 18; Candow 1989, 27, 29.

10. Royal Commission on Seals and the Sealing Industry in Canada 1986, 23.

11. Fisheries and Oceans Canada, "The Harp Seal," at http://www.dfo-mpo.gc.ca/; Horwood 1960, 39.

12. Candow 1989, 16, 30, 34–36.

13. Horwood 1960, 39.

14. Coish 1979, 25–27.

15. Horwood 1960, 39.

16. In 1895, sealers began to record harp seals separately from the other kinds of seals they landed, allowing a more accurate calculation of total harp seal catches. Figures after this time are therefore lower than the recorded *total* seal catch figures from earlier times.

17. Candow 1989, 45, 47; Fisheries and Oceans Canada, "The Harp Seal," at http://www.dfo-mpo.gc.ca. For a description of sealing during World War I and the Great Depression, see also Coish 1979, 42–50.

18. Kean 1935, 131.

19. Candow 1989, 47, 58, 169, 174.

20. Of the 8 vessels that left Norway for the Front in 1939, 1 sank en route and 2 were lost at the Front. The 5 vessels to return to Norway took a catch of 33,000 seals—about one-third of the catch of the Newfoundland vessels.

21. Candow 1989, 113, 155; Fisheries and Oceans Canada 2003, 6.

Chapter 20

1. Lillie 1955, 244, as quoted in Candow 1989, 116.

2. Horwood 1960, 40–41.

3. Royal Commission on Seals and the Sealing Industry in Canada 1986, 66–67.

4. See Serge Deyglun's letter to the editor in Lust 1967, in front matter.

5. Fisheries and Oceans Canada 2003, 6; Candow 1989, 109. In the final decade before the ban on whitecoats, furs accounted for 95 percent of sealskin profits, whereas leather accounted for just 5 percent (Candow 1989, 175).

6. Coish 1979, 258. For a personal account of the anti-sealing campaign, see Davies 1989.

7. The Canadian government eventually agreed in 1976 to authorize the use of the Norwegian *hakapik*—similar to a gaff, but having a slightly bent, blunt-tipped metal pick instead of a hook at one end.

8. After the Canadian government banned large vessels (over 65 feet in length) from the Gulf in 1972, the anti-sealing protests shifted from the Gulf to the Front. The protestors also began to focus more on the role of the Norwegian fleet at the Front (11 Norwegian and 7 Canadian vessels went to the Front in 1973).

9. Lee 1988, 23. Because seals have a swimming reflex that can continue even after death, it's hard to tell whether a seal is dead or alive in video footage of a skinning: like a chicken, a pup may move for a while even after dying. Also, because fluid to protect their corneas flows continuously from their tear ducts, pups appear to be weeping even when they aren't.

10. Lee 1988, 25. John Lee analyzes the power of words and metaphors to alter the "reality" of the Canadian seal hunt.

11. Weyler 2004, 352–363. Robert Hunter (1979, 248) credits Paul Watson and Walrus Oakenbough with the inspired idea of spraying the whitecoats with an indelible dye. For a more critical account of the Greenpeace anti-sealing campaign, see Harter 2004; for a more critical analysis of the anti-sealing movement as an industry, see Allen 1979.

12. See Weyler 2004, 601n6. Environmental critics of the seal hunt gained strength after *National Geographic* published an estimate of the 1975 western Atlantic harp seal population—based on a new technique to count whitecoats on the ice as black spots using ultraviolet aerial photographs—that put the whitecoat population at less than 200,000 and the total harp seal population at less than 1 million (Lavigne 1976, 137). Although the estimate now appears to have been too low, it armed critics with new "objective" data showing the seals were in crisis.

13. Weyler 2004, 360–363. See also Hunter 1979, 248–296; Brown and May 1989, 44–49; Dale 1996, 90–91.

14. Candow 1989, 124.

15. Weyler 2004, 493–501.

16. Revenue Canada revoked IFAW's tax-exempt status in 1977, arguing that the fund's efforts to end the seal hunt made it a political, not a charitable, organization. This pushed the organization to relocate the following year to Hyannis, Massachusetts, on Cape Cod.

17. Allen 1979, 426n8.

18. Fisheries and Oceans Canada 2003, 6.

19. Royal Commission on Seals and the Sealing Industry in Canada 1986, 68–69.

20. Joyce 1982, 23–24; Goar 1982, 23–25; Candow 1989, 136. The temporary import ban did not affect Greenland, a European Community member, because native sealers only hunted adults with rifles.

21. David Story, "Shooting Hold in Seal Hunt Protests," *Toronto Star*, 15 February 1984, as quoted in Lee 1988, 24.

22. Fisheries Minister Pierre De Bané, as quoted in "Ottawa Rejects Ban on Seal Hunt," *New York Times*, 10 March 1984, 2.

23. Candow 1989, 189.

24. Fisheries and Oceans Canada, "The Harp Seal," at http://www.dfo-mpo .gc.ca.

25. Canadian Broadcasting Corporation (TV), "Ottawa Ends Large-Scale Seal Hunt," 30 December 1987, clip at http://archives.cbc.ca/.

26. This "triumph" brought considerable economic hardship to some fishing communities, particularly among the Inuit. The primary campaign against the hunting of harp whitecoats caused sealskin prices to fluctuate and finally collapse worldwide. The Inuit were, in the words of John Amagualuk of the Tungavik Federation of Nunavut (quoted in Dale 1996, 91), "innocent bystanders," caught in a war where the activists, media, and consumers (especially in Europe) failed to distinguish between the Inuit hunt for adult seals and the large-scale commercial hunt for whitecoats. The income of the Inuit in Labrador, for example, fell by one-third because of lost sealing revenue.

27. Davies 1989, 220.

Chapter 21

1. Canadian Fisheries Minister Brian Tobin at press conference, 18 December 1995, as quoted in Lavigne 1995, 11n3.

2. Tobin's estimate was made during an interview on Canadian Broadcasting Corporation (TV), "The Seal Hunt Makes a Comeback," 18 December 1995, clip at http://archives.cbc.ca/.

The evidence connecting the resurgence in the population of Atlantic harp seals with collapsing cod stocks is highly debatable. Seals eat not only cod, but also predators of cod like squid. The typical diet of an adult harp seal is capelin, sand lance, arctic cod, crabs, squid, shrimp, and krill. On average, commercial cod account for only about 3 percent of a harp seal's diet. Fisheries and Oceans Canada today recognizes the complex and uncertain connections between seal and cod populations, and no longer justifies the commercial seal hunt as necessary to protect or revive cod stocks.

3. Fisheries and Oceans Canada 2003, 24.

4. Friedman 2003, 5.

5. Schultz and Barnes 2002, 56. Indirect subsidies, such as support for processing facilities, remain in place. Gary Gallon, director of the Canadian Institute for Business and the Environment, estimates that the Canadian government funneled over C$20 million into sealing from 1995 to 2002, which includes financing for local fisheries offices and Coast Guard support. Gallon's comments are summarized in Schultz and Barnes 2002, 56.

6. Fisheries and Oceans Canada 2003, 9, 24.

7. Fisheries and Oceans Canada, "Seals and Sealing in Canada: Facts about Seals, 2004–2005," and "Socio-Economic Impact of the Atlantic Coast Seal Hunt," at http://www.dfo-mpo.gc.ca/.

The 2003–2004 sealing figures are for Canada only. A hunt for Northwest Atlantic harp seals occurs in Greenland, too. Unlike Canada, Greenland does not mandate a total allowable catch. Currently, Greenland sealers land between 90,000 and 110,000 harp seals, as well as around 7,500 hooded seals. The struck-and-loss rate is high in this hunt, however, so that these sealers may actually kill two seals for every one landed.

8. In 2002, the province of Newfoundland accounted for over 90 percent of the landed catch of seals in Canada. The profits from the hunt constituted 25–35 percent of the annual income of the province's typical sealer, even though the hunt is a relatively small part of the C$600 million Newfoundland fishery.

9. Fisheries and Oceans Canada data, summarized in IFAW 2005, 10. The average landed value of a beater pelt in 2004 (C$48) was more than twice its average landed value in 1997 (C$22), when a bedlamer pelt earned on average C$15, a ragged jacket C$12, and an adult harp seal pelt C$9 (Canadian Institute for Business and the Environment 2001, 23).

10. In the mid-1990s, demand for seal penises was high in Asian markets, where they were thought to enhance sexual performance; as a result, male seal carcasses were worth at least two times more than female ones. With the launching of Viagra in 1998, demand for seal penises fell markedly. Today little demand remains.

11. Dion Dakins, as quoted in Armstrong 2006b, A3. The data in this paragraph are from Statistics Canada, International Trade Division, summarized in Armstrong 2006b.

12. See Fisheries and Oceans Canada, "Socio-economic Impact of the Atlantic Coast Seal Hunt," at http://www.dfo-mpo.gc.ca/.

13. In its five-year management plan for 2006–2010, the Canadian Department of Fisheries and Oceans adjusts the total allowable catch to reflect changing environmental conditions.

14. Canadian Fisheries Minister Loyola Hearn, as quoted in Armstrong 2006a, A8.

15. Marine Mammal Regulations, *Canada Gazette*, pt. 1, vol. 1, no. 9 (2 March 2002), 507.

16. Fisheries and Oceans Canada 2003, 1.

17. Daoust et al. 2002, 693.

18. See the Web sites http://www.protectseals.org/ and http://www.ifaw.org/. Notably, the environmental organization WWF does not oppose what it sees as a sustainable and reasonably humane seal hunt.

19. The Green Party of Canada's official position on sealing is at http://www .greenparty.ca/. The Martin Sheen quote is at "Martin Sheen Speaks Out Against the Seal Hunt," at http://www.seashepherd.org/. The Paul McCartney quote is from "Harp Seal Hunt a 'Stain' on Canada, McCartney says," CBC News, Canadian Broadcasting Corporation, at http://www.cbc.ca/. The Sea Shepherd Conservation Society quotes are at http://www.seashepherd.org/.

20. See the Humane Society of the United States, "Facts about the Canadian Seal Hunt," at http://www.hsus.org/.

21. See the Sea Shepherd Conservation Society Web site, http://www .seashepherd.org/.

22. IFAW 2005, 6, 8–9.

23. "Bad Science: Harp Seals' Future on Thin Ice," 10 March 2005, at http:// www.greenpeace.org/; Cox 2005, A15 (Bruce Cox is executive director of Greenpeace Canada). See also Johnston and Santillo 2005.

24. See "Boycott Canadian Seafood," at http://www.sealhunt.ca/.

25. Murphy 2004, B3. See also the IFAW Web site, http://www.ifaw.org/.

26. Dion Dakins, as quoted in Armstrong 2006b, A3.

Chapter 22

1. Royal Commission on Seals and the Sealing Industry in Canada 1986, 84, 96. For a description of the pro-sealing "counterprotest" in the second half of the 1970s, see Lamson 1979.

2. Bevan 2005, A11.

3. IFAW, "Who We Are," at http://www.ifaw.org/.

Chapter 23

1. Many books survey the evolution of the politics of global environmentalism over the last four decades. See, for example, Guha 2000; Maniates 2003; Lipschutz 2003; Switzer 2004; Elliott 2004; Conca and Dabelko 2004; Speth 2004; Clapp and Dauvergne 2005; Dryzek 2005; Dryzek and Schlosberg 2005; Chasek, Downie, and Brown 2006; Betsill, Hochstetler and Stevis 2006; DeSombre 2007.

2. See, for a sample of the literature on NGOs and global environmentalism, Keck and Sikkink 1998; Lee and So 1999; Newell 2000; Bryner 2001; Wapner

2002; Hochstetler 2002; Gunter 2004; Park 2005; Pellow 2007; Betsill and Corell 2007.

3. WWF 2006, 5.

4. See the Forest Stewardship Council Web site, http://www.fsc.org/. See also WWF 2006, 4. On Home Depot and the FSC, see "Wood Purchasing Policy," at http://www.corporate.homedepot.com/. For more on certification, nongovernmental forces, and changing forest practices, see Cashore, Auld and Newsom 2004; Gulbrandsen 2005, 2006; Espach 2006; Cashore et al. 2007.

5. For background on the Marine Stewardship Council, see http://www.msc .org/. See also Gulbrandsen 2005.

6. Alan Meier, Lawrence Berkeley National Laboratory in California, summarized in "Pulling the Plug on Standby Power," *Economist*, 11 March 2006, 32.

7. For a sample of analyses of the changing nature of global environmental governance, see include Vogler 2000; Newell 2003; Falkner 2003; Jasanoff and Martello 2004; Conca 2005; Dauvergne 2005; Najam, Papa, and Taiyab 2005; Dimitrov 2005; Humphreys 2006; Speth and Haas 2006.

Chapter 24

1. See Princen, Maniates, and Conca 2002, 326–328.

2. Whiteside 2006, viii, xi, 30, xiii.

3. Whiteside 2006, 39, 153.

4. Principle 7 of the UN Global Compact refers to the definition of a "precautionary approach" from the Rio Declaration on Environment and Development, which stresses the need to apply precaution according to state "capabilities" and with "cost-effective measures." The full definition in the Rio Declaration (Principle 15) reads: "In order to protect the environment, the precautionary approach shall be widely applied by States according to their capabilities. Where there are threats of serious or irreversible damage, lack of full scientific certainty shall not be used as a reason for postponing cost-effective measures to prevent environmental degradation." See Rio Declaration on Environment and Development on the UNEP Web site, http://www.unep.org/.

5. The estimates of the worldwide market share and light output of incandescent lightbulbs are by Harry Verhaar at Philips Electronics NV in the Netherlands, reported in Gandhi 2007, A3.

References

Abele, John J. 1970. Oil Industry Warned on Pollution. *New York Times*, 29 January, 51, 60.

Ahmed, Kulsum. 1995. *Technological Development and Pollution Abatement: A Study of How Enterprises Are Finding Alternatives to Chlorofluorocarbons.* Technical Paper no. 271. Washington, DC: World Bank.

Aldrich, Lester. 2006. Consumers Eat Up Organic Beef Despite Costs, Unproven Benefits. *Wall Street Journal* (Eastern edition), 12 July, B7C.

Allen, Jeremiah. 1979. Anti Sealing as an Industry. *Journal of Political Economy* 87 (April): 423–428.

Allison, Richard. 2003. Organic Out of Fashion? Depends Who You Ask. *Farmers Weekly*, 22 August, 39.

Ananthaswamy, Anil. 2001. Green Monster: Devices for Cleaning Car Exhausts are Backfiring on the Environment. *New Scientist*, no. 2277 (10 February): 18.

American Automobile Manufacturers Association (AAMA). 1993. *AAMA Motor Vehicle Facts & Figures '93.* Detroit.

American Automobile Manufacturers Association (AAMA). 1994. *AAMA Motor Vehicle Facts & Figures '94.* Detroit.

American Automobile Manufacturers Association (AAMA). 1997. *AAMA Motor Vehicle Facts & Figures '97.* Detroit.

Andersen, Stephen O., and K. Madhava Sarma. 2002. *Protecting the Ozone Layer: The United Nations History.* London: Earthscan.

Ascherio, Alberto et al. 2006. Pesticide Exposure and Risk for Parkinson's Disease. *Annals of Neurology* 60 (August): 197–203.

Ash, Mark, Janet Livezey, and Erik Dohlman. 2006. *Soybean Backgrounder.* Washington, DC: Economic Research Service, Department of Agriculture.

Armstrong, Jane. 2006a. Ottawa Raises Quota for Seal Hunt. *Globe and Mail* (Canada), 16 March, A8.

Armstrong, Jane. 2006b. It's "Not Pretty," but Seal Business Thrives. *Globe and Mail* (Canada), 15 April, A3.

Automobile Manufacturers Association (AMA). 1935. *Automobile Facts and Figures.* New York.

Automobile Manufacturers Association (AMA). 1936. *Automobile Facts and Figures.* New York.

Automobile Manufacturers Association (AMA). 1938. *Automobile Facts and Figures.* New York.

Automobile Manufacturers Association (AMA). 1944–45. *Automobile Facts and Figures.* New York.

Automobile Manufacturers Association (AMA). 1946–47. *Automobile Facts and Figures.* New York.

Automobile Manufacturers Association (AMA). 1951. *Automobile Facts and Figures.* New York.

Bäckstrand, Karin, and Eva Lövbrand. 2006. Planting Trees to Mitigate Climate Change: Contested Discourses of Ecological Modernization, Green Governmentality and Civic Environmentalism. *Global Environmental Politics* 6 (February): 50–75.

Bacon, Christopher M., V. Ernesto Méndez, Stephen R. Gliessman, David Goodman and Jonathan A. Fox, eds. 2008. *Confronting the Coffee Crisis: Fair Trade, Sustainable Livelihoods and Ecosystems in Mexico and Central America.* Cambridge, MA: MIT Press.

Bandivadekar, Anup P., Vishesh Kumar, Kenneth L. Gunter, and John W. Sutherland. 2004. A Model for Material Flows and Economic Exchanges within the U.S. Automotive Life Cycle Chain. *Journal of Manufacturing Systems* 23, no. 1: 22–29.

Barrett, James R. 1987. *Work and Community in the Jungle: Chicago's Packinghouse Workers, 1894–1922.* Urbana: University of Illinois Press.

Barry, Patrick L., and Tony Phillips. 2006. Earth's Ozone Layer: Good News and a Puzzle. *Space and Earth Science,* 26 May. http://www.physorg.com/.

Bartlett, Captain Robert A. 1929. The Sealing Saga of Newfoundland. *National Geographic* 56, no. 1 (July): 91–130.

Bellmann, Klaus, and Anshuman Khare. 1999. European Response to Issues in Recycling Car Plastics. *Technovation* 19: 721–734.

Benedick, Richard Elliot. 1998. *Ozone Diplomacy: New Directions in Safeguarding the Planet.* Cambridge, MA: Harvard University Press.

Bent, Silas. 1925. Tetraethyl Lead Fatal to Makers. *New York Times,* 22 June, 3.

Bergeson, Lynn. 2005. *Legal Lookout: EPA Seeks to Curb PBDEs.* Washington, DC: Bergeson & Campbell, P.C.

Betsill, Michelle M., and Elisabeth Corell, eds. 2007. *NGO Diplomacy: The Influence of Nongovernmental Organizations in International Environmental Negotiations.* Cambridge, MA: MIT Press.

Betsill, Michelle, Kathryn Hochstetler, and Dimitris Stevis, eds. 2006. *Palgrave Advances in International Environmental Politics.* New York: Palgrave Macmillan.

Betts, Kellyn. 2004. PBDEs and the Environmental Intervention Time Lag. *Environmental Science and Technology* 38: 386A–387A.

Bevan, David. 2005. Facts of the Seal Hunt Are Worth Noting. *Vancouver Sun,* 14 January, A11.

Bhagwati, Jagdish. 2004. *In Defense of Globalization.* Oxford: Oxford University Press.

Bhattacharya, Hrishikes. 2002. *Commercial Exploitation of Fisheries: Production, Marketing and Finance Strategies.* New York: Oxford University Press.

Bordwell, Marilyn. 2002. Jamming Culture: Adbusters' Hip Media Campaign against Consumerism. In Thomas Princen, Michael Maniates, and Ken Conca, eds., *Confronting Consumption,* 237–253. Cambridge, MA: MIT Press.

Bosch und Siemens Hausgeräte (BSH). 2004. *Environmental and Social Responsibility 2004.* Munich.

Bosch und Siemens Hausgeräte (BSH). 2006a. *BSH at a Glance 2006.* Munich.

Bosch und Siemens Hausgeräte (BSH). 2006b. *Environmental and Corporate Responsibility 2006.* Munich.

Bradsher, Keith. 2002. *High and Mighty: SUVs—The World's Most Dangerous Vehicles and How They Got That Way.* New York: Public Affairs.

Brinkler, Kelly. 2004. *CARE Report on Automotive Glass Recycling 2004 Update.* London: Consortium for Automotive Recycling. (The consortium represents the United Kingdom's main motor vehicle manufacturers, importers, and dismantlers.)

Brooke, Lindsay. 2000. *Automotive Industries* 180, no. 6 (June): 27.

Brown, Cassie. 1972. *Death on the Ice: The Great Newfoundland Sealing Disaster of 1914.* With Harold Horwood. Toronto: Doubleday Canada.

Brown, Colin G., John W. Longworth, and Scott A. Waldron, eds. 2002. *Regionalisation and Integration in China: Lessons from the Transformation of the Beef Industry.* Aldershot, UK: Ashgate.

Brown, Lester. 2006. *Plan B 2.0: Rescuing a Planet under Stress and a Civilization in Trouble.* New York: Norton.

Brown, Michael, and John May. 1989. *The Greenpeace Story.* Scarborough, ON: Prentice-Hall Canada.

Bryner, Gary C. 2001. *Gaia's Wager: Environmental Movements and the Challenge of Sustainability.* Lanham, MD: Rowman & Littlefield.

Bucheli, Marcelo. 2005. *Bananas and Business: The United Fruit Company in Colombia, 1899–2000.* New York: New York University Press.

Built Green Colorado (Home Builders Association of Metro Denver). 2006. Whirlpool Corporation Delivers Energy Efficient Appliances with Superior Results. http://www.builtgreen.org/.

Bulkeley, Harriet, and Susanne C. Moser, eds. 2007. *Responding to Climate Change: Governance and Social Action beyond Kyoto.* Special issue. *Global Environmental Politics* 7 (May).

Burros, Marian. 2002. Eating Well: The Greening of the Herd. *New York Times,* 29 May, F1.

Burros, Marian. 2006. Grass-Fed Rule Angers Farmers. *New York Times,* 26 July, F5.

Buss, Jessica. 2004. Organic Meat Sales Are Set for Struggle. *Farmers Weekly,* 13–19 August, 29.

Buss, Jessica, and Shirley Macmillan. 2004. Co-op Claims Beef and Lamb Are Poor Relations of Organic Boom. *Farmers Weekly,* 26 November–2 December, 30.

Caballero, Benjamin, and Barry M. Popkin, eds. 2002. *The Nutrition Transition: Diet and Disease in the Developing World.* Boston: Academic Press.

Cagin, Seth, and Philip Dray. 1993. *Between Earth and Sky: How CFCs Changed Our World and Endangered the Ozone Layer.* New York: Pantheon Books.

Canadell, Josep G. et al. 2007. Contributions to Accelerating Atmospheric CO_2 Growth from Economic Activity, Carbon Intensity, and Efficiency of Natural Sinks. *Proceedings of the National Academy of Sciences* 104 (November 20): 18866–18870.

Canadian Institute for Business and the Environment. 2001. *The Economics of the Canadian Sealing Industry.* Montreal: Canadian Institute for Business and the Environment.

Candow, James E. 1989. *Of Men and Seals: A History of the Newfoundland Seal Hunt.* Ottawa: Environment Canada.

Carle, Eric. 2002. *"Slowly, Slowly, Slowly," Said the Sloth.* New York: Philomel Books.

Carlson, Laurie Winn. 2001. *Cattle: An Informal Social History.* Chicago: Ivan R. Dee.

Cashore, Benjamin, Graeme Auld, and Deanna Newsom. 2004. *Governing through Markets: Forest Certification and the Emergence of Non-State Authority.* New Haven, CT: Yale University Press.

Cashore, Benjamin, Elizabeth Egan, Graeme Auld, and Deanna Newsom. 2007. Revising Theories of Non-State Market Driven (NSMD) Governance: Lessons from the Finnish Forest Certification Experience. *Global Environmental Politics* 7 (February): 1–44.

Cass, Loren. 2005. Norm Entrapment and Preference Change: The Evolution of the European Union Position on International Emissions Trading. *Global Environmental Politics* 4 (May): 38–60.

Centers for Disease Control and Prevention. 2005. *MMWR Morbidity and Mortality Weekly Report* 54, no. 20 (27 May): 513–516.

Chantraine, Pol. 1980. *The Living Ice: The Story of the Seals and the Men Who Hunt Them in the Gulf of St. Lawrence*. Translated by David Lobdell. Toronto: McClelland and Stewart.

Chasek, Pamela S., David L. Downie, and Janet Welsh Brown. 2006. *Global Environmental Politics*, 4th ed. Boulder, CO: Westview Press.

Chen, Ming. 2005. End-of-Life Vehicle Recycling in China: Now and the Future. *Journal of the Minerals, Metals, and Materials Society* 57, no. 10 (October): 20–26.

China Sourcing Reports. 2005. *Refrigerators and Freezers*. http://www.chinasourcingreports.com/.

Ciantar, Christopher, and Mark Hadfield. 2004. A Study of Tribological Durability with Associated Environmental Impacts of a Domestic Refrigerator. *Materials and Design* 25: 331–341.

Clapp, Jennifer. 1997. The Illegal CFC Trade: An Unexpected Wrinkle in the Ozone Protection Regime. *International Environmental Affairs* 9, no. 4: 259–273.

Clapp, Jennifer. 2001. *Toxic Exports: The Transfer of Hazardous Wastes from Rich to Poor Countries*. Ithaca, NY: Cornell University Press.

Clapp, Jennifer. 2002. The Distancing of Waste: Overconsumption in a Global Economy. In Thomas Princen, Michael Maniates, and Ken Conca, eds., *Confronting Consumption*, 155–176. Cambridge, MA: MIT Press.

Clapp, Jennifer, and Peter Dauvergne. 2005. *Paths to a Green World: The Political Economy of the Global Environment*. Cambridge, MA: MIT Press.

Cloud, John. 2003. Why the SUV Is All the Rage. *Time*, 24 February, 34–44.

Clover, Charles. 2004. *The End of the Line: How Overfishing Is Changing the World and What We Eat*. London: Random House.

Cohen, Maurie J., and Joseph Murphy. 2001. *Exploring Sustainable Consumption: Environmental Policy and the Social Sciences*. Oxford: Pergamon.

Coish, Calvin E. 1979. *Season of the Seal: The International Storm over Canada's Seal Hunt*. Saint John's, NF: Breakwater.

Cole, Doug. 1997. *Elliston: The Story of a Newfoundland Outport*. Portugal Cove, NF: ESPress.

Conca, Ken. 2002. Consumption and Environment in a Global Economy. In Thomas Princen, Michael Maniates, and Ken Conca, eds., *Confronting Consumption*, 133–153. Cambridge, MA: MIT Press.

Conca, Ken. 2005. *Governing Water: Contentious Transnational Politics and Global Institution Building*. Cambridge, MA: MIT Press.

Conca, Ken, and Geoffrey D. Dabelko, eds. 2004. *Green Planet Blues: Environmental Politics from Stockholm to Johannesburg*, 3rd ed. Boulder, CO: Westview Press.

Cone, Marla. 2005. *Silent Snow*. New York: Grove Press.

Cooper, Tim. 2005. Slower Consumption: Reflections on Product Life Spans and the "Throwaway Society." *Journal of Industrial Ecology* 9, nos. 1–2 (Winter–Spring): 51–67.

Cox, Bruce. 2005. A Precautionary Tale. *Ottawa Citizen*, 10 March, A15.

Crandall, J. R., K. S. Bhalla, and N. J. Madeley. 2002. Designing Road Vehicles for Pedestrian Protection. *British Medical Journal* 324 (11 May): 1145–1148.

Crocker, David A., and Toby Linden, eds. 1998. *Ethics of Consumption: The Good Life, Justice, and Global Stewardship.* Lanham, MD: Rowman & Littlefield.

Dale, Stephen. 1996. *McLuhan's Children: The Greenpeace Message and the Media.* Toronto: Between the Lines.

Dalton, Rex. 2005. Satellite Tags Give Fresh Angle to Tuna Quota. *Nature* 434 (28 April): 1056.

Daly, Herman E. 1993. The Perils of Free Trade. *Scientific American*, November, 50–57.

Daly, Herman E. 1996. *Beyond Growth: The Economics of Sustainable Development.* Boston: Beacon.

Daly, Herman E. 2005. Economics in a Full World. *Scientific American*, September, 100–107.

Daoust, Pierre-Yves, Alice Cook, Trent K. Bollinger, Keith G. Campbell, and James Wong. 2002. Animal Welfare and the Harp Seal Hunt in Atlantic Canada. *Canadian Veterinary Journal* 43 (September): 687–694.

Dauvergne, Peter. 1997. *Shadows in the Forest: Japan and the Politics of Timber in Southeast Asia.* Cambridge, MA: MIT Press.

Dauvergne, Peter. 2001. *Loggers and Degradation in the Asia-Pacific: Corporations and Environmental Management.* Cambridge: Cambridge University Press.

Dauvergne, Peter, ed. 2005. *Handbook of Global Environmental Politics.* Cheltenham, UK: Edward Elgar.

Davies, Brian. 1989. *Red Ice: My Fight to Save the Seals.* London: Methuen.

Davidson, Cliff I., ed. 1999. *Clean Hands: Clair Patterson's Crusade Against Environmental Lead Contamination.* Commack, NY: Nova Science.

Davis, Devra Lee, and Pamela S. Webster. 2002. The Social Context of Science: Cancer and the Environment. *Annals of the American Academy* 584, no. 1: 13–34.

Depledge, Joanna. 2007. A Special Relationship: Chairpersons and the Secretariat in the Climate Change Negotiations. *Global Environmental Politics* 7 (February): 45–68.

DeSombre, Elizabeth R. 2007. *The Global Environment and World Politics*, 2nd ed. London: Continuum International.

Diamond, Jared M. 2005. *Collapse: How Societies Choose to Fail or Succeed.* New York: Viking.

Dimitrov, Radoslav. 2005. *Science and International Environmental Policy: Regimes and Nonregimes in Global Governance*. Lanham, MD: Rowman & Littlefield.

Downie, David Leonard, and Terry Fenge, eds. 2003. *Northern Lights against POPs: Combatting Toxic Threats in the Arctic*. Montreal: McGill-Queen's University Press.

Dressler, Andrew E., and Edward A. Parson. 2006. *The Science and Politics of Global Climate Change*. Cambridge: Cambridge University Press.

Dryzek, John. 2005. *Politics of the Earth: Environmental Discourses*, 2nd ed. Oxford: Oxford University Press.

Dryzek, John, and David Schlosberg, eds. 2005. *Debating the Earth: The Environmental Politics Reader*, 2nd ed. Oxford: Oxford University Press.

Dunn, James A., Jr. 1998. *Driving Forces: The Automobile, Its Enemies, and the Politics of Mobility*. Washington, DC: Brookings Institution.

Eby, Clare Virginia, ed. 2003. *The Jungle: Upton Sinclair. An Authoritative Text: Contexts and Backgrounds. Criticism*. New York: Norton.

Ecology Center. 2005. *Moving Towards Sustainable Plastics: A Report Card on the Six Leading Automakers*. Ann Arbor, MI: Ecology Center.

Economist Intelligence Unit. 2005. World Energy Outlook: Driven by Demand. *Energy and Electricity Outlook*, June. http://www.eiu.com/.

Ehrlich, Paul R., and Anne H. Ehrlich. 2004. *One With Nineveh: Politics, Consumption, and the Human Future*. Washington, DC: Island Press.

El-Fadel, Mutasem, and Zaher Hashisho. 2001. Phase-Out of Leaded Gasoline in Developing Countries: Approaches and Prospects for Lebanon. *Journal of Environmental Assessment Policy and Management* 3 (March): 35–59.

Electrolux. 2005. *Our World, Our Approach: Sustainability Report 2005*. Stockholm.

Elliott, Lorraine. 2004. *The Global Politics of the Environment*, 2nd ed. New York: New York University Press.

Energy Information Administration. 2005. *International Energy Outlook*. July. Washington, DC: Department of Energy.

Energy Information Administration. 2007. *Emissions of Greenhouse Gases in the United States 2006*. November. Washington, DC: Department of Energy.

ENERGY STAR. 2006. *ENERGY STAR: Overview of 2005 Achievements*. Washington, DC: ENERGY STAR and Environmental Protection Agency.

Espach, Ralph. 2006. When is Sustainable Forestry Sustainable? The Forest Stewardship Council in Argentina and Brazil. *Global Environmental Politics* 6 (May): 55–84.

Euromonitor International. 2003. *Global Market Information Database: The World Market for Electrical Appliances*. November. http://www.euromonitor.com/.

Falkner, Robert. 2003. Private Environmental Governance and International Relations: Exploring the Links. In David Humphreys, Matthew Paterson, and Lloyd Pettiford, eds., *Global Environmental Governance for the Twenty-First Century: Theoretical Approaches and Normative Considerations*. Special issue. *Global Environmental Politics* 3 (May): 72–87.

Fenton, Michael D. 2000. Iron and Steel Scrap. In *U.S. Geological Survey Minerals Yearbook—1999*, 40.1–40.5. Washington, DC: Department of the Interior, Geological Survey.

Filho, Kepler Euclides. 2004. Supply Chain Approach to Sustainable Beef Production from a Brazilian Perspective. *Livestock Production Science* 90, no. 1: 53–61.

Fisheries and Oceans Canada. 2003. *Atlantic Seal Hunt 2003–2005: Management Plan*. Ottawa.

Flegal, A. Russell. 1998. Introduction: Clair Patterson's Influence on Environmental Research. *Environmental Research* 78, no. 2: 65–70.

Fonda, Daren. 2004. The Shrinking SUV. *Time*, 30 August, 65.

Food and Agriculture Organization (FAO). 2005a. *Global Forest Resources Assessment 2005: Progress Towards Sustainable Forest Management*. Rome.

Food and Agriculture Organization (FAO). 2005b. Incentives to Curb Deforestation Needed to Counter Climate Change. FAO Media Release, 9 December 2005.

Foster, Mark S. 1981. *From Streetcar to Superhighway: American City Planners and Urban Transportation, 1900–1940*. Philadelphia: Temple University Press.

Freese, Barbara. 2003. *Coal: A Human History*. Cambridge, MA: Perseus.

Freund, Peter, and George Martin. 1993. *The Ecology of the Automobile*. Montreal: Black Rose Books.

Friedman, Thomas L. 2005. *The World Is Flat: A Brief History of the Twenty-First Century*. New York: Farrar, Straus and Giroux.

Friedman, Vanessa. 2003. Fur in a Not-So-Cold Climate: After Such a Loud Campaign against It, Why Is Fur Making a Comeback? *Financial Times*, 13 December, 5.

Frumkin, Howard. 2002. Urban Sprawl and Public Health. *Public Health Reports* 117 (May–June): 201–217.

Fuchs, Doris A., and Sylvia Lorek. 2005. Sustainable Consumption Governance: A History of Promises and Failures. *Journal of Consumer Policy* 28: 261–288.

Gallagher, Kelly Sims. 2006. *China Shifts Gears: Automakers, Oil, Pollution, and Development*. Cambridge, MA: MIT Press.

Gandhi, Unnati. 2007. Global Enlightenment Turns Off the Bulb. *Globe and Mail* (Canada), 5 March, A3.

Garcia-Johnson, Ronie. 2000. *Exporting Environmentalism: U.S. Multinational Chemical Corporations in Brazil and Mexico*. Cambridge, MA: MIT Press.

Gibney, Ling-yee. 1975. Chlorofluorocarbon-Ozone Issue Flares Up Again. *Chemical and Engineering News*, 11 August, 13.

Giddens, Anthony. 2002. *Runaway World: How Globalization Is Reshaping Our Lives.* Rev. ed. London: Profile.

Gill, Alexandra. 2002. Free-Range and Grass-Fed: These Cows Are Over the Moon. *Globe and Mail* (Canada), 15 January, L8.

Gladwell, Malcolm. 2000. *The Tipping Point: How Little Things Can Make a Big Difference.* Boston: Little, Brown.

Goar, Carol. 1982. Seal Wars: The Final Battle? *Maclean's*, 6 December, 23–25.

Gold, Mark. 2004. *The Global Benefits of Eating Less Meat.* Petersfield, Hampshire, UK: Compassion in World Farming Trust.

Goldenberg, Suzanne. 2004. Taboo: Unscrupulous: Culture of Indifference Leaves America Open to BSE. *Guardian* (London), 12 January, 13.

Goodland, Robert. 1998. The Case against the Consumption of Grain-Fed Meat. In David A. Crocker and Toby Linden, eds., *Ethics of Consumption: The Good Life, Justice, and Global Stewardship*, 95–112. Lanham, MD: Rowman & Littlefield.

Goodwin, Neva R., Frank Ackerman, and David Kiron, eds. 1997. *The Consumer Society.* Washington, DC: Island Press.

Gordon, Bruce, Richard Mackay, and Eva Rehfuess. 2004. *Inheriting the World: The Atlas of Children's Health and the Environment.* Geneva: World Health Organization.

Gould, Larry. 1962. Albemarle Paper Surprises Wall St. *New York Times*, 23 September, 1, 29.

Gouveia, Lourdes, and Arunas Juska. 2002. Taming Nature, Taming Workers: Constructing the Separation between Meat Consumption and Meat Production in the U.S. *Sociologia Ruralis* 42 (October): 370–390.

Graebner, William. 1987. Hegemony through Science: Information Engineering and Lead Toxicology, 1925–1965. In David Rosner and Gerald Markowitz, eds., *Dying for Work: Workers' Safety and Health in Twentieth-Century America*, 140–159. Bloomington: Indiana University Press.

Green, Rhys E., Stephen J. Cornell, Jörn P. W. Scharlemann, and Andrew Balmford. 2005. Farming and the Fate of Wild Nature. *Science*, 28 January, 550–555.

Greenberg, Nadivah. 2006. Shop Right: American Conservatisms, Consumption, and the Environment. *Global Environmental Politics* 6 (May): 85–111.

Gross, Daniel (and the editors at *Forbes* magazine). 1996. Henry Ford and the Model T. In *Forbes: Greatest Business Stories of All Time*, 75–89. New York: Wiley.

Gudoshnikov, Sergey, Lindsay Jolly, and Donald Spence, eds. 2004. *The World Sugar Market.* Cambridge: Woodhead.

Guha, Ramachandra. 2000. *Environmentalism: A Global History*. New York: Longman.

Gulbrandsen, Lars H. 2005. Mark of Sustainability? Challenges for Fishery and Forestry Eco-Labeling. *Environment* 47 (June): 8–23.

Gulbrandsen, Lars H. 2006. Creating Markets for Eco-Labelling: Are Consumers Insignificant? *International Journal of Consumer Studies* 30 (September): 477–489.

Gunter, Michael M. 2004. *Building the Next Ark: How NGOs Protect Biodiversity*. Dartmouth, NH: Dartmouth College Press and University Press of New England.

Haegi, Marcel. 2002. A New Deal for Road Crash Victims. *British Medical Journal* 324 (11 May): 1110.

Hamer, Mick. 1996. A Hundred Years of Carnage. *New Scientist*, no. 2042 (10 August): 14–15.

Hammar, Henrik, and Åsa Löfgren. 2004. Leaded Gasoline in Europe: Differences in Timing and Taxes. In Winston Harrington, Richard D. Morgenstern, and Thomas Sterner, eds., *Choosing Environmental Policy: Comparing Instruments and Outcomes in the United States and Europe*, 192–205. Washington, DC: Resources for the Future.

Hammitt, James K. 2004. CFCs: A Look Across Two Continents. In Winston Harrington, Richard D. Morgenstern, and Thomas Sterner, eds., *Choosing Environmental Policy: Comparing Instruments and Outcomes in the United States and Europe*, 158–174. Washington, DC: Resources for the Future.

Harrington, Winston, Richard D. Morgenstern, and Thomas Sterner, eds. 2004. *Choosing Environmental Policy: Comparing Instruments and Outcomes in the United States and Europe*. Washington, DC: Resources for the Future.

Harter, John-Henry. 2004. Environmental Justice for Whom? Class, New Social Movements, and the Environment: A Case Study of Greenpeace Canada, 1971–2000. *Labour/Le Travail*, no. 54 (Fall): 83–121.

Heazle, Michael. 2006. *Scientific Uncertainty and the Politics of Whaling*. Seattle: University of Washington Press.

Helleiner, Eric. 2002. Think Globally, Transact Locally: The Local Currency Movement and Green Political Economy. In Thomas Princen, Michael Maniates, and Ken Conca, eds., *Confronting Consumption*, 255–274. Cambridge, MA: MIT Press.

Hendrickson, Mary, and William Heffernan. 2005. Concentration of Agricultural Markets. University of Missouri, Columbia, Department of Rural Sociology. http://www.foodcircles.missouri.edu/CRJanuary05.pdf.

Higgins, Amy. 2001. New Fridges: More Features, Less Juice. *Machine Design*, 5 April, 50–58.

Hilton, F. Hank. 2001. Later Abatement, Faster Abatement: Evidence and Explanations From the Global Phaseout of Leaded Gasoline. *Journal of Environment and Development* 10, no. 3 (September): 246–265.

Hirst, Paul, and Grahame Thompson. 1999. *Globalization in Question: The International Economy and the Possibilities of Governance*, 2nd ed. Cambridge, MA: Polity Press.

Hites, Ronald A. et al. 2004. Global Assessment of Polybrominated Diphenyl Ethers in Farmed and Wild Salmon. *Environmental Science and Technology* 38, no. 19: 4945–4949.

Hochstetler, Kathryn. 2002. After the Boomerang: Environmental Movements and Politics in the La Plata River Basin. *Global Environmental Politics* 2 (November): 35–57.

Hond, Frank den, Peter Groenewegen, and Nico M. van Straalen, eds. 2003. *Pesticides: Problems, Improvements, Alternatives*. Oxford: Blackwell Science.

Horwood, Harold. 1960. Tragedy on the Whelping Ice. *Canadian Audubon* 22, no. 2 (March–April): 37–41.

Hough, Peter. 1998. *The Global Politics of Pesticides: Forging Consensus from Conflicting Interests*. London: Earthscan.

Humphreys, David. 2006. *Logjam: Deforestation and the Crisis of Global Governance*. London: Earthscan.

Humphreys, David, Matthew Paterson, and Lloyd Pettiford, eds. 2003. *Global Environmental Governance for the Twenty-First Century: Theoretical Approaches and Normative Considerations*. Special issue. *Global Environmental Politics* 3 (May).

Hunter, Robert. 1979. *Warriors of the Rainbow: A Chronicle of the Greenpeace Movement*. New York: Holt, Rinehart and Winston.

Intergovernmental Panel on Climate Change (IPCC). 2001. *Climate Change 2001: The Scientific Basis*. Geneva.

International Energy Agency. 2006. *Oil, Gas, Coal and Electricity—Quarterly Statistics*, no. 2. Paris: Organization for Economic Cooperation and Development.

International Fund for Animal Welfare (IFAW). 2005. *Seals and Sealing in Canada*. Guelph, ON.

Jackson, Kenneth T. 1985. *Crabgrass Frontier: The Suburbanization of the United States*. Oxford: Oxford University Press.

Jackson, Richard, and Glenn Banks. 2003. *In Search of the Serpent's Skin: The Story of the Porgera Gold Mine*. Brisbane, Australia: Boolarong Press.

Japanese Forestry Agency. 1993. *Forestry White Paper: Fiscal Year 1992, Summary*. Tokyo: Government of Japan.

Jasanoff, Sheila, and Marybeth Long Martello, eds. 2004. *Earthly Politics: Local and Global in Environmental Politics*. Cambridge, MA: MIT Press.

Johansen, Bruce E. 2003. *The Dirty Dozen: Toxic Chemicals and the Earth's Future*. Westport, CT: Praeger.

Johnston, Paul, and David Santillo. 2005. *The Canadian Seal Hunt: No Management and No Plan*. Amsterdam: Greenpeace International. March.

Jones, Darryl. 2003. *The Pig Sector*. Paris: OECD.

Joyce, Randolph. 1982. More than One Way to Skin a Seal Hunt. *Maclean's*, 22 March, 23–24.

Kaimowitz, David et al. 2004. Hamburger Connection Fuels Amazon Destruction, Cattle Ranching and Deforestation in Brazil's Amazon. Center for International Forestry (CIFOR). Bogor, Indonesia. April.

Kauffman, George B. 1989. Midgley: Saint or Serpent? *Chemtech*, December, 716–725.

Kean, Abram. 1935. *Old and Young Ahead: A Millionaire in Seals Being the Life History of Captain Abram Kean, O.B.E.* London: Heath Cranton.

Keck, Margaret E., and Kathryn Sikkink. 1998. *Activists Beyond Borders: Advocacy Networks in International Politics*. Ithaca, NY: Cornell University Press.

Kelly, Katy. 2005. Birth of the Cool. *U.S. News and World Report*, 15 August, 52.

Kelly, Kevin, and Heidi Dawley. 1995. Don't Get the Lead Out. *Business Week*, 15 May, 50.

Kim, Hyung Chul, Gregory A. Keoleian, and Yuhta A. Horie. 2006. *Energy Policy* 34, no. 15 (October): 2310–2323.

Kitman, Jamie Lincoln. 2000. The Secret History of Lead. *Nation*, 20 March, 11–44.

Kovarik, William. 1999. Charles F. Kettering and the 1921 Discovery of Tetraethyl Lead in the Context of Technological Alternatives. Revised version of paper presented to the Society of Automotive Engineers Conference, 1994, Baltimore. http://www.radford.edu/~wkovarik/papers/.

Kovarik, William. 2003. Ethyl: The 1920s Environmental Conflict over Leaded Gasoline and Alternative Fuels. Paper presented to the American Society for Environmental History, Annual Conference, 26–30 March, Providence, RI. http://www.radford.edu/~wkovarik/papers/.

Kovarik, William. 2005. Ethyl-leaded Gasoline: How a Classic Occupational Disease Became an International Public Health Disaster. *International Journal of Occupational and Environmental Health* 11 (October): 384–397.

Lacey, Marc. 2004a. Belatedly, Africa Is Converting to Lead-Free Gasoline. *The New York Times* (International edition), 31 October 2004, N3.

Lacey, Marc. 2004b. Sub-Saharan Africa Shifts to Lead-Free Gas. *International Herald Tribune*, 2 November, 5.

Lamson, Cynthia. 1979. *"Bloody Decks and a Bumper Crop": The Rhetoric of Sealing Counter-Protest*. Social and Economic Studies, no. 24. Saint John's, NF: Institute of Social and Economic Research, Memorial University.

Langley, Billy C. 1994. *Refrigerant Management: The Recovery, Recycling, and Reclaiming of CFCs*. Albany, NY: Delmar.

Lansdown, Richard, and William Yule. 1986. *Lead Toxicity: History and Environmental Impact*. Baltimore: Johns Hopkins University Press.

Lawrence, Felicity. 2006. Whether You Know It or Not, You'll Probably be Eating Soya Today. *Guardian* (London), 25 July, 8.

Lavigne, David M. 1976. Life or Death for the Harp Seal. *National Geographic* 149, no. 1 (January): 129–142.

Lavigne, David M. 1995. Seals and Fisheries, Science and Politics. Paper presented to the 11th Biennial Conference on the Biology of Marine Mammals, 14–18 December, Orlando, FL.

Lee, John Alan. 1988. Seals, Wolves and Words: Loaded Language in Environmental Controversy. *Alternatives* 15 (November–December): 20–29.

Lee, Yok-Shiu F., and Alvin Y. So., eds. 1999. *Asia's Environmental Movements: Comparative Perspectives.* Armonk, NY: M. E. Sharpe.

Leigh, Charles. 2005. Food for Thought as Organic Sales Grow: More Expensive but Experts Split on Whether It Is Any Better for You. *Guardian* (London), 31 May 2005, 18.

Lemonick, Michael D., and David Bjerklie. 2004. How We Grew So BIG. *Time*, 7 June, 58–69.

Lewis, Jack. 1985. Lead Poisoning: A Historical Perspective. *EPA Journal*, May. http://www.epa.gov/history/topics/perspect/lead.htm.

Lichtenberg, Judith. 1998. Consuming Because Others Consume. In David A. Crocker and Toby Linden, eds. 1998. *Ethics of Consumption: The Good Life, Justice, and Global Stewardship*, 155–175. Lanham, MD: Rowman & Littlefield.

Lillie, Harry R. 1955. *The Path Through Penguin City.* London: Ernest Benn.

Lipschutz, Ronnie D. 2003. *Global Environmental Politics: Power, Perspectives, and Practice.* Washington, DC: CQ Press.

Litfin, Karen T. 1994. *Ozone Discourse: Science and Politics in Global Environmental Cooperation.* New York: Columbia University Press.

Litfin, Karen T. 2000. Environment, Wealth and Authority: Global Climate Change and Emerging Modes of Legitimation. *International Studies Review* 2 (Summer): 119–148.

Livestock and the Environment: Meeting the Challenge. 1999. Report of a study coordinated by the United Nations Food and Agriculture Organization, the United States Agency for International Development, and the World Bank. Rome: FAO.

Löfgren, Åsa, and Hammar, Henrik. 2000. The Phase Out of Leaded Gasoline in the EU Countries—A Successful Failure? Transportation Research Part D. *Transport and Environment* 5D, no. 6: 419–431.

Lomborg, Bjørn. 2001. *The Skeptical Environmentalist.* Cambridge: Cambridge University Press.

Luban, David. 1998. The Political Economy of Consumption. In David A. Crocker and Toby Linden, eds. 1998. *Ethics of Consumption: The Good Life, Justice, and Global Stewardship*, 113–130. Lanham, MD: Rowman & Littlefield.

Lust, Peter. 1967. *The Last Seal Pup: The Story of Canada's Seal Hunt.* Montreal: Harvest House.

Lynam, Donald R., Lillian G. Piantanida, and Jerome F. Cole, eds. 1981. (International Lead Zinc Research Organization.) *Environmental Lead.* New York: Academic Press.

MacFarlane, Alan, and Iris MacFarlane. 2004. *The Empire of Tea: The Remarkable History of the Plant that Took Over the World.* New York: Overlook Press.

MacNeill, Jim, Pieter Winsemius, and Taizo Yakushiji. 1991. *Beyond Interdependence: The Meshing of the World's Economy and the Earth's Ecology.* New York: Oxford University Press.

Madsen, Robert A. 1995. Of Oil and Rainforests: Using Commodity Cartels to Conserve Depletable Natural Resources. *International Environmental Affairs: A Journal for Research and Policy* 7 (Summer): 207–234.

Maguire, Steve, and Cynthia Hardy. 2006. The Emergence of New Global Institutions: A Discursive Perspective. *Organization Studies* 27, no. 1: 7–29.

Majtenyi, Cathy. 2005. Leaded Fuel to be Phased Out in Sub-Saharan Africa. *Voice of America,* 27 December.

Malthus, Thomas R. 1798. *An Essay on the Principle of Population.* 1st ed. London: J. Johnson. http://www.econlib.org/library/Malthus/malPop.html.

Maniates, Michael F. 2002a. Individualization: Plant a Tree, Buy a Bike, Save the World? In Thomas Princen, Michael Maniates, and Ken Conca, eds., *Confronting Consumption,* 43–66. Cambridge, MA: MIT Press.

Maniates, Michael F. 2002b. In Search of Consumptive Resistance: The Voluntary Simplicity Movement. In Thomas Princen, Michael Maniates, and Ken Conca, eds., *Confronting Consumption,* 199–236. Cambridge, MA: MIT Press.

Maniates, Michael F., ed. 2003. *Encountering Global Environmental Politics: Teaching, Learning and Empowering Knowledge.* Lanham, MD: Rowman and Littlefield.

Manno, Jack. 2002. Commoditization: Consumption Efficiency and an Economy of Care and Connection. In Thomas Princen, Michael Maniates, and Ken Conca, eds., *Confronting Consumption,* 67–99. Cambridge, MA: MIT Press.

Markowitz, Gerald, and David Rosner. 2002. *Deceit and Denial: The Deadly Politics of Industrial Pollution.* Berkeley: University of California Press.

McAuley, John W. 2003. Global Sustainability and Key Needs in Future Automotive Design. *Environmental Science and Technology* 37, no. 23: 5414–5416.

McKay, George. 1996. *Senseless Acts of Beauty: Cultures of Resistance since the Sixties.* London: Verso.

McKendrick, Neil, John Brewer, and J. H. Plumb. 1982. *The Birth of a Consumer Society: The Commercialization of Eighteenth-Century England.* Bloomington: University of Indiana Press.

McShane, Clay. 1994. *Down the Asphalt Path: The Automobile and the American City.* New York: Columbia University Press.

Meacham, Cory J. 1997. *How the Tiger Lost its Stripes: An Exploration into the Endangerment of a Species.* New York: Harcourt Brace.

Meikle, James. 2006. Farmers and Butchers Illegally Selling Meat as Organic. *Guardian* (London), 15 May, 6.

Midgley, Thomas, IV. 2001. *From the Periodic Table to Production: The Life of Thomas Midgley, Jr., the Inventor of Ethyl Gasoline and Freon Refrigerants.* Corona, CA: Stargazer.

Millstone, Erik, and Tim Lang. 2003. *The Penguin Atlas of Food: Who Eats What, Where, and Why.* London: Penguin Books.

Mittelstaedt, Martin. 2006a. Coming to Terms with Perils of Non-Stick Products. *Globe and Mail* (Canada), 29 May, A1, A8.

Mittelstaedt, Martin. 2006b. Are Plastic Products Coated in Peril? *Globe and Mail* (Canada), 31 May, A3.

Mittelstaedt, Martin. 2006c. Want a Full-Time Job? Live Chemical-Free. *Globe and Mail* (Canada), 1 June, A8.

Molina, Mario J., and F. Sherwood Rowland. 1974. Stratospheric Sink for Chlorofluoromethanes: Chlorine Atom-Catalysed Destruction of Ozone. *Nature* 249 (28 June): 810–812.

Moran, Susan. 2006. The Range Gets Crowded for Natural Beef. *New York Times*, 10 June, C1.

Moss, Michael, Richard A. Oppel Jr., and Simon Romero. 2004. Mad Cow Forces Beef Industry to Change Course. *New York Times*, 5 January, A1.

Motor Vehicle Manufacturers Association (MVMA) of the United States. 1972. 1977. 1978. 1982. 1987. *Automobile Facts & Figures.* (Title varies slightly from year to year.) Detroit.

Mowat, Farley. 1973. *Wake of the Great Sealers.* Drawings and pictures by David Blackwood. Boston: Little, Brown.

Moxham, Roy. 2003. *Tea: Addiction, Exploitation and Empire.* London: Constable and Robinson.

Murphy, Sean P. 2004. In Hyannis, a Quiet Force: Group Pursues Global Mission of Animal Welfare. *Boston Globe*, 15 August, B3.

Mushita, Andrew T., and Carol B. Thompson. 2002. Patenting Biodiversity? Rejecting WTO/TRIPS in Southern Africa. *Global Environmental Politics* 2 (February): 65–82.

Myers, Norman, and Jennifer Kent. 2004. *The New Consumers: The Influence of Affluence on the Environment.* Washington, DC: Island Press.

Myers, Ransom A., and Boris Worm. 2003. Rapid Worldwide Depletion of Predatory Fish Communities. *Nature* 423 (15 May): 280–283.

Nader, Ralph. 1965. *Unsafe at any Speed: The Designed-in Dangers of the American Automobile.* New York: Grossman.

Najam, Adil, Mihaela Papa, and Nadaa Taiyab. 2005. *Global Environmental Governance: A Reform Agenda.* Winnipeg: International Institute for Sustainable Development.

Nantulya, Vinand M., and Michael R. Reich. 2002. The Neglected Epidemic: Road Traffic Injuries in Developing Countries. *British Medical Journal* 324 (11 May): 1139–1141.

National Automobile Chamber of Commerce. 1931. *Facts and Figures of the Automobile Industry.* New York: National Automobile Chamber of Commerce.

Needleman, Herbert L. 1998. Clair Patterson and Robert Kehoe: Two Views of Lead Toxicity. *Environmental Research* 78, no. 2: 79–85.

Needleman, Herbert L. 2000. The Removal of Lead from Gasoline: Historical and Personal Reflections. *Environmental Research* 84 (September): 20–35.

Needleman, Hebert L. et al. 1979. Deficits in Psychologic and Classroom Performance of Children with Elevated Dentine Lead Levels. *New England Journal of Medicine* 300, issue 13: 689–695.

Newell, Peter. 2000. *Climate for Change: Non-State Actors and the Global Politics of the Greenhouse.* Cambridge: Cambridge University Press.

Newell, Peter. 2003. Globalization and the Governance of Biotechnology. In David Humphreys, Matthew Paterson, and Lloyd Pettiford, eds., *Global Environmental Governance for the Twenty-First Century: Theoretical Approaches and Normative Considerations.* Special issue. *Global Environmental Politics* 3 (May): 56–71.

Newell, Richard G., and Kristian Rogers. 2004. Leaded Gasoline in the United States. In Winston Harrington, Richard D. Morgenstern, and Thomas Sterner, eds., *Choosing Environmental Policy: Comparing Instruments and Outcomes in the United States and Europe,* 174–191. Washington, DC: Resources for the Future.

Nierenberg, Danielle. 2005. *Happier Meals: Rethinking the Global Meat Industry.* Worldwatch Paper no. 171. Washington, DC: Worldwatch Institute.

Noble, Kenneth B. 1984. Lead Industry Digs in Its Heels on Gas Additive. *New York Times,* 6 August, E8.

Nriagu, Jerome O. 1998. Clair Patterson and Robert Kehoe's Paradigm of "Show Me the Data" on Environmental Lead Poisoning. *Environmental Research* 78, no. 2: 71–78.

Oliveira, Victor, and David Davis. 2006. *Recent Trends and Economic Issues in the WIC Infant Formula Rebate Program.* Economic Research Report no. ERR-22. Washington, DC: Department of Agriculture.

O'Neill, Kate. 2000. *Waste Trading among Rich Nations: Building a New Theory of Environmental Regulation.* Cambridge, MA: MIT.

O'Rourke, Dara. 2005. Market Movements: Nongovernmental Organization Strategies to Influence Global Production and Consumption. *Journal of Industrial Ecology* 9, nos. 1–2 (Winter–Spring): 115–128.

Ostro, Bart. 2000. *Lead: Evaluation of Current California Air Quality Standards with Respect to Protection of Children.* Report prepared for California Air Resources Board and California Office of Environmental Health Hazard Assessment. http://www.oehha.ca.gov/air/criteria_pollutants/AQAC2.html.

Park, Susan. 2005. How Transnational Environmental Advocacy Networks Socialize International Financial Institutions: A Case Study of the International Finance Corporation. *Global Environmental Politics* 5 (November): 95–119.

Parson, Edward A. 2003. *Protecting the Ozone Layer: Science and Strategy.* Oxford: Oxford University Press.

Paterson, Matthew. 2000. Car Culture and Global Environmental Politics. *Review of International Studies* 26 (April): 253–270.

Paterson, Matthew. 2007. *Automobile Politics: Ecology and Cultural Political Economy.* Cambridge: Cambridge University Press.

Patterson, Clair C. 1965. Contaminated and Natural Lead Environments of Man. *Archives of Environmental Health* 11 (September): 344–360.

Pearce, David, ed. 1990. *Elephants, Economics and Ivory.* London: Earthscan.

Pellow, David. 2007. *Resisting Global Toxics: Transnational Movements for Environmental Justice.* Cambridge, MA: MIT Press.

Phiri, Patson. 2006. Slow Progress on Phase Out of Leaded Fuel in SADC. *All Africa*, 17 January.

Pollan, Michael. 2002. Power Steer. *New York Times Magazine*, 31 March, sec. 6, pp., 44–51, 68, 71–72, 76–77.

Popkin, Barry M. 1993. Nutritional Patterns and Transitions. *Population and Development Review* 19 (March), 138–157.

Porter, Gareth. 1999. Trade Competition and Pollution Standards: "Race to the Bottom" or "Stuck at the Bottom"? *Journal of Environment and Development* 8, no. 2: 133–151.

Porter, Richard C. 1999. *Economics at the Wheel: The Costs of Cars and Drivers.* San Diego, CA: Academic Press.

Pounds, J. Alan, and Robert Puschendorf. 2004. Ecology: Clouded Futures. *Nature* 427 (8 January): 107–109.

Pretty, Jules N. 2005. *The Pesticide Detox: Towards a More Sustainable Agriculture.* London: Earthscan.

Princen, Thomas. 2002a. Consumption and Its Externalities: Where Economy Meets Ecology. In Thomas Princen, Michael Maniates, and Ken Conca, eds., *Confronting Consumption*, 23–42. Cambridge, MA: MIT Press.

Princen, Thomas. 2002b. Distancing: Consumption and the Severing of Feedback. In Thomas Princen, Michael Maniates, and Ken Conca, eds., *Confronting Consumption*, 103–131. Cambridge, MA: MIT Press.

Princen, Thomas. 2005. *The Logic of Sufficiency.* Cambridge, MA: MIT Press.

Princen, Thomas, Michael Maniates, and Ken Conca, eds. 2002. *Confronting Consumption.* Cambridge, MA: MIT Press.

Purdum, Todd S. 2005. High Priest of the Pasture. *New York Times*, 1 May, sec. 6, p. 76.

Raloff, Janet. 2003. New PCBs? Throughout Life, Our Bodies Accumulate Flame Retardants, and Scientists Are Beginning to Worry. *Science News* 164, no. 17: 266–268.

Read, Nicholas. 2007. Burger Eaters and Cows Large Part of Problem. *Vancouver Sun*, 10 February, C7.

Redclift, Michael. 1996. *Wasted: Counting the Costs of Global Consumption.* London: Earthscan.

Renner, Rebecca. 2004. Tracking the Dirty Byproducts of a World Trying to Stay Clean. *Science*, 10 December, 1887.

Rixson, Derrick. 2000. *The History of Meat Trading.* Nottingham: Nottingham University Press.

Robbins, Jim. 2003. Balancing Cattle, Land and Ledgers. *New York Times*, 8 October, F8.

Robert, Joseph C. 1983. *Ethyl: A History of the Corporation and the People Who Made It.* Charlottesville: University of Virginia Press.

Robertson, Roland. 1992. *Globalization: Social Theory and Global Culture.* London: Sage.

Robertson, Roland, and Kathleen M. White. 2002. *Globality and Modernity.* London: Sage.

Robinson, Nick. 2000. *The Politics of Agenda Setting: The Car and the Shaping of Public Policy.* Aldershot, UK: Ashgate.

Rogers, Ben. 2003. *Beef and Liberty.* London: Chatto and Windus.

Roosevelt, Margot. 2006. Grass-Fed Revolution. *Time*, 12 June, 76–78.

Rosenblatt, Roger. 1999. *Consuming Desires: Consumption, Culture, and the Pursuit of Happiness.* Washington, DC: Island Press.

Rosner, David, and Gerald Markowitz. 1985. A "Gift of God"? The Public Health Controversy over Leaded Gasoline during the 1920s. *American Journal of Public Health* 75 (April): 344–352.

Rowland, Sherwood. 2000. Ozone Hole. In Heather Newbold, ed., *Life Stories: World-Renowned Scientists Reflect on Their Lives and the Future of Life on Earth*, 134–155. Berkeley: University of California Press.

Rowlands, Ian H. 2000. Beauty and the Beast? BP's and Exxon's Positions on Global Climate Change. *Environment and Planning C: Government and Policy* 18, no. 3: 339–354.

Royal Commission on Seals and the Sealing Industry in Canada. 1986. *Seals and Sealing in Canada: Report of the Royal Commission.* Vol. 2. Montreal.

Rudel, Ruthann A. et al. 2007. Animal Mammary Gland Carcinogens. *Cancer* 109 (15 June): 2635–2667.

Rutledge, Margie. 2005. Cutting through the Smog. *Globe and Mail* (Canada), 30 July, F7.

Sancton, Thomas et al. 2001. Life without Beef. *Time Europe*, 26 February, 18–23.

Sayeg, Philip. 1998. *Successful Conversion to Unleaded Gasoline in Thailand*. World Bank Technical Paper no. 410. Washington, DC: World Bank.

Schaffer, Paul. 2004. The Fuss over Fluff. *American Metal Market* 112, no. 7 (16 February): 4–5.

Schecter, Arnold et al. 2006. Polybrominated Diphenyl Ether (PBDE) Levels in an Expanded Market Basket Survey of U.S. Food and Estimated PBDE Dietary Intake by Age and Sex. *Environmental Health Perspectives* 114 (October): 1515–1520.

Schlosser, Eric. 2001. *Fast Food Nation: The Dark Side of the All-American Meal*. Boston: Houghton Mifflin.

Schlosser, Eric. 2002. The Killing Zone. *Guardian* (London), 23 February, 26.

Schmitz, Andrew, ed. 2002. *Sugar and Related Sweetener Markets: International Perspectives*. New York: CABI.

Scholte, Jan Aart. 2005. *Globalization: A Critical Introduction*, 2nd ed. Basingstoke, UK: Palgrave Macmillan.

Schor, Juliet B. 2004. *Born to Buy: The Commercialized Child and the New Consumer Culture*. New York: Scribner.

Schor, Juliet B. 2005. Sustainable Consumption and Worktime Reduction. *Journal of Industrial Ecology* 9, nos. 1–2 (Winter–Spring): 37–50.

Schor, Juliet B., and Douglas B. Holt, eds. 2000. *The Consumer Society Reader*. New York: New Press.

Schultz, Stacey, and Julian E. Barnes. 2002. How Canadian Subsidies Led to a New Slaughter of Seal Pups. *U.S. News and World Report*, 6 May, 56.

Selin, Henrik, and Noelle Eckley. 2003. Science, Politics, and Persistent Organic Pollutants: Scientific Assessments and their Role in International Environmental Negotiations. *International Environmental Agreements: Politics, Law and Economics* 3, no. 1: 17–42.

Selin, Henrik, and Stacy D. VanDeveer. 2006. Global Standards: Hazardous Substances and E-Waste Management in the European Union. *Environment* 48, no. 10 (December): 6–17.

Severson, Kim. 2005. Give 'Em a Chance, Steers Will Eat Grass. *New York Times*, 1 June, F1.

Shapiro, Irving S. 1975. The Ozone Layer vs. the Aerosol Industry: Du Pont Wants to See Them Both Survive. *Washington Post*, 9 July, 13.

Sheehan, Molly O'Meara. 2001. *City Limits: Putting the Brakes on Sprawl*. Washington, DC: Worldwatch Institute.

Simms, Ciaran, and Desmond O'Neill. 2005. Sports Utility Vehicles and Older Pedestrians: A Damaging Collision. *British Medical Journal* 331 (8 October): 787–788.

Sinclair, Stuart. 1983. *The World Car: The Future of the Automobile Industry*. London: Euromonitor.

Sinclair, Upton. 1906. *The Jungle*. New York: Doubleday, Page.

Sinclair, Upton. 1962. *The Autobiography of Upton Sinclair*. New York: Harcourt, Brace, & World.

Skaer, Mark. 2001. The Decade of Refrigerant Chaos—and Change. *Air Conditioning, Heating and Refrigeration News* 212, no. 18: 106–111.

Skaggs, Jimmy. 1986. *Prime Cut: Livestock Raising and Meatpacking in the United States 1607–1983*. College Station: Texas A & M Press.

Skjæreth, Jon Birger, and Tora Skodvin. 2001. Climate Change and the Oil Industry: Common Problems, Different Strategies. *Global Environmental Politics* 1 (November): 18–42.

Smil, Vaclav. 2000. *Feeding the World: A Challenge for the Twenty-First Century*. Cambridge, MA: MIT Press.

Smil, Vaclav. 2002. Eating Meat: Evolution, Patterns, and Consequences. *Population and Development Review* 28 (December): 599–639.

Smith, George Otis. 1920. Where the World Gets Its Oil: But Where Will Our Children Get It When American Wells Cease to Flow? *National Geographic Magazine* 37, no. 2 (February): 181–202.

Soluri, John. 2006. *Banana Cultures: Agriculture, Consumption, and Environmental Change in Honduras and the United States*. Austin: University of Texas Press.

Southerton, Dale, Heather Chappells, and Bas Van Vliet, eds. 2004. *Sustainable Consumption: The Implications of Changing Infrastructures of Provision*. Cheltenham, UK: Edward Elgar.

Speth, James Gustave. 2004. *Red Sky at Morning: America and the Crisis of the Global Environment*. New Haven, CT: Yale University Press.

Speth, James Gustave, and Peter M. Haas. 2006. *Global Environmental Governance*. Washington, DC: Island Press.

Steinberg, Paul F. 2001. *Environmental Leadership in Developing Countries: Transnational Relations and Biodiversity Policy in Costa Rica and Bolivia*. Cambridge, MA: MIT Press.

Stern, Nicholas. 2006. *Stern Review on the Economics of Climate Change*. London: H.M. Treasury.

Stevenson, Drury. 2005. No Purchase Necessary. *Cornell International Law Journal* 38, no. 1: 251–262.

Stoett, Peter J. 1997. *The International Politics of Whaling*. Vancouver: University of British Columbia Press.

Stokstad, Eric. 2006. Du Pont Settlement to Fund Test of Potential Toxics. *Science*, 6 January, 26–27.

Striffler, Steve. 2002. *In the Shadows of State and Capital: The United Fruit Company, Popular Struggle, and Agrarian Restructuring in Ecuador, 1900–1995*. Durham, NC: Duke University Press.

Striffler, Steve, and Mark Moberg, eds. 2003. *Banana Wars: Power, Production, and History in the Americas*. Durham, NC: Duke University Press.

Subak, Susan. 1999. Global Environmental Costs of Beef Production. *Ecological Economics* 30, no. 1: 79–91.

Switzer, Jacqueline Vaughn. 2004. *Environmental Politics: Domestic and Global Dimensions*. Belmont, CA: Thomson/Wadsworth.

Talbot, John M. 2004. *Grounds for Agreement: The Political Economy of the Coffee Commodity Chain*. Boulder, CO: Rowman & Littlefield.

Tedlow, Richard S. 1990. *New and Improved: The Story of Mass Marketing in America*. New York: Basic Books.

Thévenot, Roger. 1979. *A History of Refrigeration throughout the World*. Translated by J. C. Fidler. Paris: International Institute of Refrigeration.

Thomas, Chris et al. 2004. Extinction Risk from Climate Change. *Nature* 427 (8 January): 145–148.

Thomas, Daniel A. 2004. Safety First. *Planning* 70 (May): 26–29.

Thomson, Vivian E. 2000. Getting the Lead Out: Grab-Bag Ethics in Environmental Policy Making. In Joel Reichart and Patricia H. Werhane, eds., *Environmental Challenges to Business*, 185–203, Ruffin Series in Business Ethics, vol. 2. St. Joseph, MN: Society for Business Ethics.

Timberg, Craig. 2006. Era of Leaded Gas Comes to an End in Most of Africa. *Washington Post*, 1 January, A14.

Tucker, Richard P. 2002. Environmentally Damaging Consumption: The Impact of American Markets on Tropical Ecosystems in the Twentieth Century. In Thomas Princen, Michael Maniates, and Ken Conca, eds., *Confronting Consumption*, 177–195. Cambridge, MA: MIT Press.

Tucker, Richard P., and J. F. Richards, eds. 1983. *Global Deforestation and the Nineteenth-Century World Economy*. Durham, NC: Duke University Press.

United Nations Centre for Human Settlements (UNCHS). 2001. *The State of the World's Cities*. Nairobi.

United Nations Conference on Trade and Development (UNCTAD). 2001. *World Investment Report 2001: Promoting Linkages*. New York.

United Nations Conference on Trade and Development (UNCTAD). 2002. *World Investment Report 2002: Transnational Corporations and Export Competitiveness*. World Investment Report 2002. New York. (See also the data from the 1990, 1995, and 2000 *Reports*).

United Nations Conference on Trade and Development (UNCTAD). 2007. *World Investment Report 2007: Transnational Corporations, Extractive Industries and Development*. World Investment Report 2007. New York.

United Nations Environment Programme (UNEP). 1989. *CFCs for Aerosols, Sterilants and Miscellaneous Uses*. Report by the Technical Options Committee on Aerosols, Sterilants and Miscellaneous Uses. Nairobi.

United Nations Environment Programme (UNEP). 2002. *GEO: Global Environment Outlook 3: Past, Present and Future Perspectives.* Nairobi.

United Nations Environment Programme (UNEP). 2004. Leaded-Petrol Phase Out in Sub-Saharan Africa. UNEP news release, 7 May.

United Nations Environment Programme (UNEP). 2005a. Basic Facts and Data on the Science and Politics of Ozone Protection. UNEP news release, November.

United Nations Environment Programme (UNEP). 2005b. Sub-Saharan Africa Celebrates Leaded Petrol Phase-Out. News release 2005/66, 27 December.

United Nations Population Fund (UNFPA). 2004. *State of the World Population 2004.* New York.

United Nations Secretariat. 2001. *World Population Prospects: The 2000 Revision 2001* (Highlights). New York: Population Division of the Department of Economic and Social Affairs. New York.

United States Environmental Protection Agency (EPA). 1973. *EPA's Position on the Health Implications of Airborne Lead.* Washington, DC.

United States Public Health Service. 1925. Proceedings of a Conference to Determine Whether or Not there is a Public Health Question in the Manufacture, Distribution or Use of Tetraethyl Lead Gasoline. *Public Health Bulletin*, no. 158. Washington, DC.

United States Surgeon General. 1926. The Use of Tetraethyl Lead and Its Relation to Public Health. *Public Health Bulletin*, no. 163. Washington, DC.

Victor, David G. 2006. Toward Effective International Cooperation on Climate Change: Numbers, Interests and Institutions. *Global Environmental Politics* 6 (August): 90–103.

Vince, Gaia. 2004. Dust Storms on the Rise Globally. NewScientist.com news service, 20 August.

Vogel, David. 1995. *Trading Up: Consumer and Environmental Regulation in a Global Economy.* Cambridge, MA: Harvard University Press.

Vogel, David. 2003. The Hare and the Tortoise Revisited: The New Politics of Consumer and Environmental Regulation in Europe. *British Journal of Political Science* 33, no. 1: 557–580.

Vogel, David. 2004. Trade and Environment in the Global Economy: Contrasting European and American Perspectives. In Norman Vig and Michael Faure, eds., *Green Giants? Environmental Policies of the United States and the European Union*, 231–252. Cambridge, MA: MIT Press.

Vogler, John. 2000. *The Global Commons: Environmental and Technological Governance*, 2nd ed. New York: Wiley.

Wakefield, Julie. 2002. The Lead Effect? *Environmental Health Perspectives* 110 (October): A574–A580.

Wall, Derek. 1999. *Earth First! and the Anti-Roads Movement.* London: Routledge.

Walter, Katey M., Laurence C. Smith, and F. Stuart Chapin III. 2007. Methane Bubbling from Northern Lakes: Present and Future Contributions to the Global Methane Budget. *Philosophical Transactions of the Royal Society A* 365 (15 July): 1657–1676.

Wapner, Paul. 2002. Horizontal Politics: Transnational Environmental Activism and Global Cultural Change. *Global Environmental Politics* 2 (May): 37–62.

Wapner, Paul, and John Willoughby. 2005. The Irony of Environmentalism: The Ecological Futility but Political Necessity of Lifestyle Change. *Ethics and International Affairs* 19 (Fall): 77–89.

Ward's Communications. 2002. *Ward's Motor Vehicle Facts & Figures: Documenting the Performance and Impact of the U.S. Auto Industry.* Southfield, MI.

Ward's Communications. 2003. *Ward's Motor Vehicle Facts & Figures: Documenting the Performance and Impact of the U.S. Auto Industry.* Southfield, MI.

Warner, Melanie. 2006. U.S. Restaurant Chains Find There Is No Too Much. *New York Times*, 28 July, C5.

Weatherhead, Elizabeth C., and Signe Bech Andersen. 2006. The Search for Signs of Recovery of the Ozone Layer. *Nature* 441 (4 May): 39–45.

Weiser, Rivka. 2005. *Teflon and Human Health: Do the Charges Stick? Assessing the Safety of PFOA.* Edited by Gilbert L. Ross. New York: American Council on Science and Health.

Weisskopf, Michael. 1988. CFCs: Rise and Fall of Chemical "Miracle"; Chlorofluorocarbons vs. Ozone. *Washington Post*, 10 April, A1.

Wenz, Peter S. 2001. *Environmental Ethics Today.* Oxford: Oxford University Press.

Westra, Laura, and Patricia H. Werhane, eds. 1998. *The Business of Consumption: Environmental Ethics and the Global Economy.* Lanham, MD: Rowman & Littlefield.

Weyler, Rex. 2004. *Greenpeace: How a Group of Ecologists, Journalists and Visionaries Changed the World.* Vancouver, BC: Raincoast Books.

Whirlpool Corporation. 2006. *Whirlpool Annual Report.* Benton Harbor, MI.

Whiteside, Kerry H. 2006. *Precautionary Politics: Principle and Practice in Confronting Environmental Risk.* Cambridge, MA: MIT Press.

Wild, Antony. 2004. *Coffee: A Dark History.* London: Fourth Estate.

Wilkinson, Todd. 2003. Organic Beef Gains Amid Mad Cow Scare. *Christian Science Monitor*, 29 December, 3.

Wolf, Martin. 2004. *Why Globalization Works.* New Haven, CT: Yale University Press.

Womack, James P., Daniel T. Jones, and Daniel Roos. 1990. *The Machine That Changed the World.* New York: Rawson Associates.

Woollard, Robert F., and Aleck S. Ostry, eds. 2000. *Fatal Consumption: Rethinking Sustainable Development.* Vancouver: University of British Columbia Press.

World Bank. 2005. *World Development Indicators.* Washington, DC.

World Bank. 2006. *Global Economic Prospects 2007: Managing the Next Wave of Globalization.* Washington, DC.

World Commission on Environment and Development. 1987. *Our Common Future.* Oxford: Oxford University Press.

World Health Organization (WHO) and World Bank. 2004. *World Report on Road Traffic Injury and Prevention.* Geneva.

World Health Professions Alliance. 2001. Antimicrobial Resistance: World Health Professions Alliance Fact Sheet. http://www.whpa.org/factresistance.htm. (Includes news release "Antibiotic Resistance is a Global Public Health Threat Calling for Urgent International Action," 26 March 2001.)

World Meteorological Organization. 2006. *Greenhouse Gas Bulletin*, no. 1 (March).

World Resources Institute. 2005. *Ecosystems and Human Well-Being: Synthesis.* Washington, DC.

Worldwatch Institute. 2003. *State of the World 2003.* Washington, DC.

Worldwatch Institute. 2004a. *Good Stuff? A Behind-the-Scenes Guide to the Things We Buy.* Washington, DC.

Worldwatch Institute. 2004b. *State of the World 2004: The Consumer Society.* Washington, DC.

Worldwatch Institute. 2006. *Vital Signs 2006–2007.* Washington, DC.

World Wildlife Fund / World Wide Fund for Nature (WWF). 2006. *WWF Highlights*, no. 4 (June): 1–7.

Worm, Boris et al. 2006. Impacts of Biodiversity Loss on Ocean Ecosystem Services. *Science*, 3 November, 787–790.

Yago, Glenn. 1984. *The Decline of Transit: Urban Transportation in German and U.S. Cities, 1900–1970.* Cambridge: Cambridge University Press.

Yang, Xiaohua. 1995. Globalization of the Automobile Industry: The United States, Japan, and the People's Republic of China. Westport, CT: Praeger.

Yardley, Jim. 2004. Chinese Take Recklessly to Cars (Just Count the Wrecks). *New York Times* (International edition), 12 March, A4.

Zabarenko, Deborah. 2006. 2005 Was Warmest Year on Record: NASA. *Reuters*, 24 January.

Zhao, Jimin. 2005. Implementing International Environmental Treaties in Developing Countries: China's Compliance with the Montreal Protocol. *Global Environmental Politics* 5 (February): 58–81.

Zhao, Jimin, and Leonard Ortolano. 1999. Implementing the Montreal Protocol in China: Use of Cleaner Technology in Two Industrial Sectors. *Environmental Impact Assessment Review* 19, nos. 5–6 (September–November): 499–519.

Zhao, Jimin, and Leonard Ortolano. 2003. The Chinese Government's Role in Implementing Multilateral Environmental Agreements: The Case of the Montreal Protocol. *China Quarterly* 175 (September): 708–725.

Zhao, Jimin, and Qing Xia. 1999. China's Environmental Labeling Program. *Environmental Impact Assessment Review* 19, nos. 5–6 (September–November): 477–497.

Zulu, Vincent. 1999. Chilly Reception: Mixed Feelings about Europe's Roaming Fridges. *New Internationalist*, no. 310 (March): 6.

Index